D0095015

Federal Bodysnatchers and the New Guinea Virus

Also by Robert S. Desowitz

New Guinea Tapeworms and Jewish Grandmothers

The Thorn in the Starfish

The Malaria Capers

Who Gave Pinta to the Santa Maria?

Federal Bodysnatchers and the New Guinea Virus

PEOPLE, PARASITES, POLITICS

Robert S. Desowitz

W. W. NORTON & COMPANY
NEW YORK   LONDON

Printed in the United States of America
First Edition

The text of this book is composed in New Baskerville
with the display set in Tekton
Composition by Sue Carlson
Manufacturing by Quebecor World, Fairfield
Book design by Mary A. Wirth
Production manager: Amanda Morrison

LIBRARY OF CONGRESS CATALOGING-IN-PUBLICATION DATA
Desowitz, Robert S.
Federal bodysnatchers and the new guinea virus : people, parasites, politics / by Robert S. Desowitz.
p. cm.
Includes index.
ISBN 0-393-05185-4
1. Communicable diseases—Popular works. 2. Social medicine—Popular works. 3. Epidemiology—Popular works. I. Title.

RA643 .D47 2002
616.9—dc21 2002071431

W. W. Norton & Company, Inc., 500 Fifth Avenue, New York, N.Y. 10110
www.wwnorton.com

W. W. Norton & Company Ltd., Castle House, 75/76 Wells Street, London W1T 3QT

1 2 3 4 5 6 7 8 9 0

As always, for my wife, Carrolee

and the F_1 generation
Duba and Gregory

and the F_2 generation in order of appearance
Robert, Brandon, Michael, and Zachary

Contents

Housekeeping Prologue

Edwin Barber has been my editor through five books and twenty years. He is a cherished friend; I trust his judgement and in this prologue I am heeding his advice that I explain to the readers what may seem to be omissions—if not downright eccentricities.

Shortly before Christmas 2001, Ed phoned me and said "anthrax!" It was the publisher's opinion that a "public" book on infectious diseases, such as this one, should, in keeping with the disturbing events, have one or more chapters on germ warfare–bioterrorism. Specifically, there should be an analysis of anthrax as a microbial weapon. I disagreed, a rare act of rebellion. The book was finished; the books were closed. Only copy editing and galley checking for typos remained before publication. Also, I was about to be resurrected from retirement to go "onstage" again for a semester as visiting professor at the University of Texas–Houston School of Public Health. On further (rationalizing) reflection, it is more insightful into how—and how well—the biomedical establishment responds to new microbial threats to use examples such as the West Nile virus whose scenario has been played out to a conclusion. As of this writing, the anthrax story is far from known.

The other bit of housekeeping is to explain why you won't find a bibliography of references, or even of "selected reading" in this book. And here I am more uncertain if I made the right call. Publishing an article in a peer-reviewed science journal demands extensive citation of the studies that were the foundation of the submitted report. The Literature Cited section of the paper allows the reviewer and reader to check on the author's methodologies, interpretation, and innovativeness. Believe me, the author of a science book for the general readership has also read and consulted the many hundreds of published papers and documents that are the substance of his or her book. The "donkey work" has already been done and I do not believe that the popular book should be cluttered with long lists of citations. Very few people read or refer to them. It is also a different world today, where the Internet "library" provides almost instantaneous information on new publications and other, less formal, sources.

Perhaps these are, as the British used to say "bolshy" decisions. But this book is "bolshy," confronting some cherished and established practices of the infectious disease research industry. Finally, unless there is a stunning future epiphany, I consider this to be my final book. Retirement, age, and the disconnection from "hands-on" research in field and laboratory—the wellspring of much of my writing—brings part of my life to a close. It closes with the same note of irritation that I have expressed these past twenty years with the discontinuity between research and reality in dealing with some of the greatest health problems of humankind. Writing has been a learning experience for me, but most important, it has been a dialogue with my readers—all of whom I consider to be my friends and colleagues. So to all of you—thank you and farewell.

ROBERT S. DESOWITZ
May 2002

Federal Bodysnatchers and the New Guinea Virus

Introduction

Water under My Grandmother's Bridge

By 1370 the Black Death had killed one-half of the English people. The microscopic bacterium *Yersinia pestis* would forever change England's social and economic systems. When the plague first infiltrated from the European continent, in about 1346, England lived under rigid feudalism. Except for the king, God's agent on earth, everybody belonged to someone else—the lords to the king, the knights to the lords, the small landowner or land renter to the knights, and at the bottom of the heap, making the whole system work, were the miserable peasants, bound to their master, with no hope that they or their children could ever own the land they worked.

The plague was an equal opportunity killer; knowing no class distinction, it struck down noble and serf alike. But there were more peasants than lords; it was these serfs and villeins, the surviving work force, who derived a new power and freedom as the great estates fell into shambles. It became a laborer's seller's market, and they simply left their bondage to seek hire from anyone who would pay the highest wage. The formerly impoverished serf accumulated wealth from the now impoverished petty nobility and bought or rented the land. Thus, the Black Plague created the English petite bourgeoisie. Feudalism died, the seeds of democracy had been sown.

There were other, economic, consequences. Before the plague, England had been a wealthy, grain-exporting country. During and after the plague, without the labor force needed for intensive agriculture, the lands were converted to pasture for sheep. Baa, baa black sheep—England had become a wool industry nation (actually, the nursery rhyme was a lament on the downturn of the wool economy).

Seven centuries later, in a golden age of science, the human animal is again beset by a great, implacable, uncontrollable killing plague. It may also be that when, in some future century, historians come to consider "our" AIDS plague they will focus on the social-economic consequences rather than its biomedical nature. They would comment on how AIDS prompted the militancy of gays followed by their social-economic enfranchisement. The attack by the Africans on the intellectual property/patent rights of the multinational pharmaceutical industry and how this changed worldwide marketing and pricing of therapies will be discussed by the legal and social historians. The social scientists would emphasize how (a) AIDS caused a woman's revolt and empowered them or (b) women remained passive even when threatened by the certain death of AIDS, and how they tolerated male sexual behavior that could only be considered criminal—as it would be if a man came upon a woman and injected her with a syringe-full of a lethal pathogen. The same psychoanalysts (if they still exist in future centuries) would marvel that that old sexual drive out-powered the instinctual death fear. The microbiologists of that future time would write about their appreciation of the AIDS virus and how it had taught them the ways of our immune system. The infectious disease specialist may speak of how AIDS revealed other, unknown or neglected diseases. If today's conditions remain unchanged, the future public health expert might lament science's great failure to discover a vaccine or cure despite decades of research and funding second only to that for cancer. All, whatever their expertise, will look back at the Plague of AIDS as the virus that changed the century.

AIDS appeared approximately twenty years ago and almost innocently we didn't realize what its social impact was to be or how it would affect the other major infectious diseases. We were so wrong when AIDS first came to our countrymen and I was to write in my "public" immunity book, *The Thorn in the Starfish* (W. W. Norton, 1987): "The present (1986) guesstimate is that about 20 percent of the "healthy" serological positive will ultimately develop AIDS." The world's "real" health problems were the perpetual, infectious pathogens, some of which became more pathogenic in the HIV-infected. It was those "other" parasites that were the topics of my first book, *New Guinea Tapeworms and Jewish Grandmothers* (W. W. Norton, 1981).

It's my Jewish grandmother's and her tapeworm's twentieth anniversary and a lot of medical biology has gone under the bridge since 1981. In its two-decade flow, the river of medical science has been both profound and turbulent. It was the end of an era of heroic-sized water impoundment projects in the tropics. The central theme of *NGTJG* was that these man-made environmental changes were intensifying the parasitic, bacterial, and viral infectious diseases. For people of the tropics, particularly Africans, it was a time of mixed hope and disillusionment. The cold war was at its hottest and money poured into the third world from capitalists and communists alike. Donor governments had allocated funds for health, education, roads—the infrastructure and services meant to bring these poor nations into the capitalist/communist good life of the late twentieth century. In Africa this largesse hardly reached its intended purposes. It was skimmed, if a 90 percent theft can be called "a skim," by rapacious political leaders. Joseph Mobutu of the Congo alone "skimmed" about $5 billion.

Nevertheless, there was hope in the villages that some good medicine would come from the American or Russian largesse. The villagers I met in Africa, Asia, and the Pacific knew that their health problems were mostly the old familiars—malaria, diarrheas, and respiratory infections. And they knew what they wanted: curative drugs for these infections and preventative vaccines to save their children.

To the infectious diseases-public health clinicians, the view from the village was not so simple in 1980. If anything, the health of the third world was at a twenty-five-year low and was about to sink much lower. True, smallpox had been globally eradicated by a triumphant immunization campaign. But the eradication of malaria campaign had failed, a failure that left behind insecticide-resistant mosquitoes and drug-resistant parasites like garbage in a ship's wake. Research and development for the infectious diseases health problems of the third world had essentially been abandoned. Pharmaceutical companies, no longer beholden to colonial governments, had stopped trying to find new drugs or vaccines for diseases peculiar to the tropics. The rapidly skyrocketing costs of bringing a drug to market with the heavy tolls to be paid for research, patent lawyers, clinical investigation, regulatory approval, and advertising had made tropical medicines unacceptably unprofitable. For university and institutional-based scientists, drug research was nonfunded, unglamourous, and professionally unrewarding.

Vaccine research was faring somewhat better with a bold new search for an immunization against malaria. However, by 1990 there was still no vaccine for human use (nor is there one now) and the initiative had become tarnished by the theft of large amounts of money from the Agency for International Development's malaria vaccine program. As for new insecticides to combat malaria, dengue, yellow fever, plague, typhus, and filariasis, that desire of discovery had been annihilated by the Silent Spring.

So what's new? Well, we are a lot smarter than we were in 1981. We can manipulate genes and exploit the molecular biology underlying the mechanisms of disease and immunity. We can make designer drugs and numerous biomedical cures. Even so, we are, globally, in many respects much sicker. AIDS marches through the world untreated and for many, untreatable. The malaria mantra of latter-day malariologists is 3 billion at risk, 300 million infected and 3 million dead as it marches through the tropical world untreated and, for too many, untreatable. New pathogens appear in old places; old pathogens appear in new places. Old infectious diseases in old places flourish and are ignored.

Still, looking backward to the eighties, my New Guinea Jewish Grandmother, with all her health problems, was living in her fool's paradise. Most pathogens could be treated with effective and affordable drugs. Fevers and diarrheas were known old enemies that could be diagnosed and, in many instances, public health strategies could be implemented. Biomedical scientists on the public payroll worked for the public good without the distraction of the legal license for self-enrichment. Global warming, and its possible medical impact, was but a blip on the health horizon. Bioterrorism was the topic of science fiction and secret consultation of the covert national services. Pathogen-carrying immigrants were just beginning to flood into the industrial north. There were diseases that the industrial north had never heard of, never considered. West Nile virus fever was only in the tropical medicine textbooks and maybe a microbiology graduate student would read about it as insurance for an exam. Cryptosporidium was studied by a few offbeat protozoologists. No water purifying system would tout the product as making the drinking water safe from Cryptosporidium. Ebola what? A bingo-oid game like tombola? Are we, the top macroscopic animal, still in bondage to the microscopic world of pathogenic creatures?

Chapter

1

West Nile-on-the-Danube: The Virus before 1999

If "emerging diseases" had a sense of humor, they would be amused at being "discovered" like some lost tribe. They've always been there; only circumstance has newly brought them to the body and mind of their human associates. And so it is with the West Nile virus.

The human episode of the West Nile story began in 1937 in the West Nile District of northern Uganda. A killing sleeping sickness epidemic (trypanosomiasis) was then at its killing apogee and the Sleeping Sickness Service was screening natives for this tsetse fly-transmitted infection. One of those caught in the surveillance net was a thirty-seven-year-old woman from a village called Omogo. She did not test positive for the trypanosome but she did have a temperature of 100.6°F and a blood sample was taken. The physician who examined her, Dr. A. W. Burke, noted rather testily that she refused to admit feeling ill, probably because she didn't want to be hospitalized. Her blood sample was, nevertheless, sent to the Yellow Fever Research Institute in Entebbe.

In 1937, yellow fever was still a power in the tropical world. It had retreated from the temperate zone where, for example, in 1793, it had killed a tenth or more of the population of Philadelphia. However, it remained entrenched in Africa and the tropical Americas. In Africa, yellow fever cut a wide swath from the west coast to central Africa and partially penetrated east Africa where it was transmitted to humans and monkeys by a variety of mosquitoes. The British had established the Yellow Fever Research Institute in order to be alert to any new viral intrusion into their east African colonies. Of course, in 1937, no one had actually *seen* a yellow fever virus—or any other virus for that matter. They were too small to be seen by even the best optical microscopes and their visualization had to await the introduction of the electron microscope.*

The viruses were unseen but not unknown. Filters of ceramic and colloidan membranes with porosities known to exclude even the smallest bacteria had been made since the time of Louis Pasteur. Filtrates were inoculated into experimental animals and, later, into tissue cell cultures. Any consequent pathogenic effects were observed. Passage of material from animal to animal, or culture to culture, with

* The conventional optical microscope's maximum magnification is limited by factors of light waves, refractive indices, and the configuration of the glass lenses. Between 1924 and 1926 the French physicist Louis de Broglie calculated that the wavelength of the electron stream emitted by the newly invented cathode ray tube would allow a magnification vastly greater than that from light waves when controlled by voltage and focused by magnetic fields. In 1931 a two-lens electron microscope was built, followed two years later by what might be considered a truly functional instrument. In 1935 the first commercially available electron microscope was built in England. According to the literature, the first "sighting" of a virus by the electron microscopists was in 1942; it was the vaccina virus of smallpox immunization.

pathogenic effects at each subpassage confirmed that something living, very small, and infectious was present in the original filtrate; these filterable agents were given the term *virus.* By 1937, immunology had grown from its infancy at the turn of the century into a lusty teenager. Antibodies had been characterized and their specificity of action demonstrated. Serum antibodies had been collected from a spectrum of naturally infected humans and animals and experimentally infected animals. Some antibodies neutralized the virus in a filtrate or otherwise infectious material, that is, they killed them. Some immune sera would neutralize a filtrate/virus at a given dilution (titer) but either have no effect or only partially neutralize another filtrate/virus at that titer. From applying a collection, a library, of these different acting serum antibodies, it became apparent that not only were viruses of different sizes but also, like all other living things, they were of different kinds. They were of different species and could be taxonomically sorted into related groups. When researchers in Entebbe inoculated the serum filtrate from the slightly febrile but not demonstrably ill lady from Omogo directly into the brains of mice, all the mice died within five to ten days. The doomed mice first became hyperactive, then weak and sluggish, then lapsed into a coma and died. Indian rhesus monkeys inoculated with the virus-infected mouse brains exhibited similar signs of encephalitis and, invariably, died. But curiously, the African *Cercopithecus* monkeys developed only a mild fever and recovered naturally, like the lady from Omogo. Two of the Entebbe researchers carrying out these animal experiments developed, when their blood was tested, very high concentrations (titers) of neutralizing antibody. They obviously had become accidentally infected during the experiments but neither showed any signs or symptoms of disease, not even a transient low fever. Thus, early in the West Nile virus story it was realized that there were marked differences in pathogenicity between hosts; mice died, rhesus monkeys died, African monkeys survived, one human (an African) had a fever but would admit to no other symptoms; and two humans (British) were completely unaffected. It was a virus, but what kind and how was it transmitted in nature? Indeed, where was it in nature?

The Entebbe virologists had serum antibody reagents that specifically neutralized yellow fever, Japanese B encephalitis, St. Louis encephalitis, and louping ill disease viruses.* The antibodies to yellow fever and St. Louis encephalitis virus had no neutralizing effect on the "Omogo" virus. Antibodies to Japanese B and louping ill viruses were partially lethal (neutralizing). From the clinical and laboratory findings, the Entebbe researchers, led by Dr. K. C. Smithburn, decided that it was a virus that had never been previously described and named it the West Nile virus. Its serological affinities to Japanese B and louping ill viruses, both known to be transmitted by blood-sucking arthropods (a mosquito and a tick, respectively), made them suspect that West Nile was also transmitted by a blood feeding, "hematophagous," to use the term fancied by science, arthropod.

About thirty million years ago, a mosquito got trapped in the sap of a tree and became immortalized in amber. The exact coevolutionary timing is not known with any certainty, but when animals developed blood the bloodsuckers that

* Louping ill disease is caused by a tick-transmitted virus and mostly affects sheep. They stagger about, "loup" as they say in Scotland, and then usually die. Humans bitten by an infected tick can also develop an encephalitis. Human infection is associated with sheep-rearing, particularly in Scotland.

fed on them appeared shortly thereafter. Then some opportunistic microbial pathogens figured out in the Darwinian way that those bloodsuckers were an efficient method of transport from one host feeding ground to another. The ancients suspected that some diseases were insect-transmitted, but proof had to await the discoveries that these infectious pathogens—viruses, bacteria, fungi, protozoa, and helminths—actually existed. The first demonstration of arthropod transmission came in 1893 when T. Smith and F. L. Kilbourne showed that the organism causing Texas Red Water fever in cattle, a *Babesia* protozoan related to malaria parasites, was carried from cow to cow by ticks. There followed, in rapid succession, the discoveries of important associations—the tsetse fly and the trypanosome by David Bruce in 1895, the mosquito and the malaria parasites by Ronald Ross in 1898, and, finally, the transmission of a virus, the mosquito and the yellow fever virus, by Walter Reed in 1900.

The electron microscope and refined serological methods were to reveal the great diversity of viruses transmitted by insects and ticks. In the progression to taxonomic, Linnaean, order they have been lumped as the acronym ARBO viruses (arthropod borne), which was split into the A and B arbovirus groups, and finally re-lumped into the family *Togaviridae* containing the genus *Flavivirus.* Some sixty species of flaviviruses are now recognized, including such notables as those causing dengue, Japanese B and St. Louis encephalides, yellow fever, and West Nile fever.

The West Nile virus looked like an ARBO virus; immunologically it quacked like an ARBO virus, but was it an ARBO virus? Three years after its taxonomic assignation in Kenya it was verified by virologists in America that West Nile was indeed transmitted by mosquitoes. C. Philip and J. Smadel, in 1943, fed *Aedes albopictus,* the "Tiger" mosquito of fearful fame, on West Nile virus-infected mice. Several weeks later these mosquitoes infected new, "clean" mice after taking their blood meal on them. M. Kitaoka in Japan carried out similar, successful transmission experiments, but he used two common Culex mosquitoes, *Culex pipiens* and *Culex tritaeniorhynchus.* Now it was understood that the West Nile virus was not only transmitted by mosquitoes but it was not very fastidious about its mosquito host; a wide variety of mosquitoes, including the Anopheles malaria vectors, could carry the infection. Later it was shown that blood-feeding ticks could also act as transmitters.

The clinical character of West Nile virus was undefined in 1950. Was its pathogenicity no more severe than a severe common cold or could it cause severe disease, even death? The first answer, or answers, that even now influence medical decision making came not from sick Africans but from cancerous, moribund New Yorkers. In the early 1950s, there was much excitement, stemming from laboratory studies, that a virus could be used therapeutically to kill cancer cells. In experimental animals with tumors—myxoma of rabbits, Ehrlich ascites tumor of mice, and mouse leukemia—certain viruses homed in on the cancerous cells and destroyed them.

Treatment for cancers was more savage and even less effective a half-century ago than it is now. Physicians were willing to try any novel approach that showed experimental promise, including viral therapy. Combating one disease with another wasn't entirely new; in 1927, the Austrian doctor Julius Wagner von Jauregg was awarded the Nobel Prize for his method of treating late-stage neurological syphilis, paresis, by inducing a malaria infection. But which virus should be used for the treatment of human cancers? The

West Nile virus was a candidate because, as far as was then known, it caused nothing more serious in humans than a flu-like illness—although all those experimental dead mice with the virus in their brains should have raised some concern. But pressing on, in 1950, two doctors from New York's Sloan-Kettering Institute of the Memorial Center for Cancer and Allied Diseases, Chester Southam and Alice Moore, began inoculating a series of ninety-five desperately ill patients with West Nile virus.

Some of these experimental patients were so terminal that they never made it to the starting line and died of their cancer within a few days of being injected with the virus, before the virus could manifest any clinical or therapeutic expression. Of the seventy-eight remaining subjects, there were, remarkably, no discernible virus-associated signs or symptoms other than a slight fever usually not exceeding 1°F above normal in sixty-nine of them. In the other eight, West Nile virus revealed its savage potential as a true pathogen. In those eight, it produced a neurologic disease variously ranging from drowsiness to coma. Nevertheless in all, recovery was complete, or near complete—even in the one person who became transiently paralyzed.

These brave patients, observed under carefully controlled hospital and laboratory conditions, established the clinical parameters of West Nile fever. The trial revealed that even in people whose immune system was weakened and whose body was ravaged by cancer the infection was usually no big thing, no more troublesome than a short bout of low fever. But the trial also showed that in a few people the virus could cause severe, albeit nonlethal, neurologic disease.

And does West Nile virus cure cancer? In six subjects, Southam and Moore reported definite tumor regression.

Unfortunately the regression was neither timely nor pronounced enough and all of the six ultimately died of their cancer. This radical experimental approach to cancer therapy was then pretty much abandoned. But not forgotten. With the new genetic technologies in hand, virologists are now working to build "smart bomb" viruses that will go directly and exclusively to the cancerous cells and destroy them. And as far as West Nile virus was concerned, at midcentury the epidemiologists and virologists considered it an African and, possibly, a European problem—a minor problem at that. America had its viral problems, but West Nile was not one of them.

Southam and Moore's clinical descriptions of experimental infections in cancer patients while fascinating and instructive, were *not* lessons from nature. The Sloan-Kettering Institute's cancer hospital was *not* the natural setting, the landscape epidemiology, of the West Nile virus. Extensive field studies were needed. Over the years, Smithburn and his associates at Kenya's Yellow Fever Research Laboratory had continued to collect and test serum samples from east and central Africa. By 1952, they had shown that it was widespread in the region, present (as evidenced by serologically positive sera) in Sudan, Uganda, and the Belgian Congo. A year later, Israeli scientists found that it was present in their young state. However, the most extensive, thorough study would come from American sailors in Egypt.

Not many Americans know of the outstanding medical research presence our Navy and Army has had, and continues to have, in the tropics. Staffed by a mixture of career military and civilian civil service employees, these units in Indonesia, Thailand, the Philippines, Malaysia, Kenya, Taiwan, and Egypt have carried out some of the most important research on tropical diseases. They have also been the "nursery" of some of today's foremost infectious disease researchers and practitioners.

One venerable and honored overseas laboratory has been the U.S. Naval Medical Research Unit No. 3 (NAMRU-3) based in Cairo. During the 1950s and 1960s, a classical era of virology, NAMRU-3's staff included a number of great virologists. One was the late Telford Work, an old friend of mine from our student days at the London School of Hygiene and Tropical Medicine. Tel combined the talents and training of physician, virologist, and ornithologist. He was also an acerbic and sometimes quite trying person (come to think of it, all the ARBO virologists I know are, brilliance aside, acerbic and quite trying persons). In 1951, Tel was at NAMRU-3 where in collaboration with the American virologist R. M. Taylor, the Egyptian virologist Faraz Rizk, and the entomologist H. G. Hurlbut, he brought his varied knowledge to bear in greatly expanding our understanding of West Nile virus epidemiology.

The Egyptian West Nile story was rather like the original scenario from Uganda—a serendipitous finding from a study on a completely different microbial pathogen. In Uganda, the trypanosome led to the West Nile virus; in Egypt, it was the poliomyelitis virus. In the summer of 1950, a NAMRU-3 scientist, Dr. John R. Paul, collected blood from 250 two- to four-year-old children living in a semirural village 15 miles north of Cairo. The intention of the survey was to determine the percentage of children that had been infected with the polio virus as evidenced by a specific antibody in their blood serum. A serum collection from a defined group is always a treasure for laboratory exploitation; polio virus was the target but the sera were also tested for antibodies to a panel of viruses, including the West Nile virus. Unexpectedly, three of the serum samples were serologically positive for West Nile. The Navy virus hunters pursued the new serological spoor, returning to the village for a full age-stratified blood collection. The laboratory results quickly revealed that a high percentage of the adults had, at some time in their lives, been infected with the virus. In the fall of 1951, the now intrigued NAMRU-3 scientists began an intensive study of the West Nile virus in the Nile River valley. Blood specimens were obtained from villagers and their domestic animals. Nonmigratory birds were captured and bled. The entomologists captured blood-sucking mosquitoes and ticks. Then they went back to the original village where the virus was originally found in young children and bled everybody of all ages. There was a special interest in the blood samples from children who had a fever of unknown origin.

The exercise was completed. Thousands of medical records had been searched and thousands of blood samples collected from man, bird, and beast. Thousands of "test tube tests" and inoculated mouse brains had been analyzed and thousands of mosquitoes and ticks processed. Finally, these brought the features of West Nile virus's face into a more distinct definition. Reassuringly, that face was not too menacing. The virus caused sickness in a few young children during the summer months; a high fever, about 102°F (39°C), lasting five or six days. During the febrile period, there was, as would be expected, sweating, malaise, and sometimes nausea or other gastrointestinal disturbances. But no deaths; no instances of neurological involvement. In a rural Egyptian village with little in the way of health resources, such fevers of childhood are almost a way of life and almost a blessing when they are of a nonfatal nature. In adults, the infection probably passed almost unnoticed. By age fifteen, over 70 percent of the villagers had the telltale specific antibodies against the virus. And to confirm West Nile's nonsevere reputation, two of NAMRU-3's laboratory technicians who had been handling infectious material became infected themselves. The course of their illness was like that in the children, nothing more.

For all its seeming harmlessness, West Nile virus held some notable surprises. First, no one had suspected how common the infection would be; three-quarters of the adults living in the Nile valley had, at some time in their lives, been infected with the virus. Second, no one had suspected how unfastidious the virus could be; it infected man, woman, beast, and bird with equal ease. Of domestic animals tested for the antibody: 86 percent of the horses; 78 percent of the camels; 72 percent of the water buffalo; and 28 percent of the sheep were positive. Birds were similarly susceptible, especially crows, who had a 63 percent positivity rate. Sparrows, pigeons, geese, ibis, and herons also commonly had the serological sign of infection.

Approximately 50,000 Egyptian mosquitoes were collected, pooled by species and ground up. The macerated mosquito tissues were filtered and the filtrate inoculated into about 1,000 baby mice in order to recover the virus from affected brains. The late, avuncular Harry Hoogstraal, NAMRU-3's longtime chief of medical zoology and the world's foremost authority on ticks, did the same for his favorite blood-sucking arthropods. Most of the isolates came from a mosquito named *Culex univittatus*, and it was concluded that this species was the main vector of West Nile virus. None of the ticks, fleas, lice, or bedbugs naturally harbored the virus but it was experimentally shown that they could biologically support it, the virus surviving and proliferating after injection into their body cavity. In 1954, after three years of study, the NAMRU-3 study was wrapped up. Following the Egyptian leads, future investigations in other places would have to attend to mosquitoes, birds, and the four-footed and two-footed hosts.

From Cairo to Club Med isn't all that far, and the West Nile virus was no stranger to French southerners. Between 1935 and 1942, some of the citizens of Montpelier, a charming university city near the Mediterranean littoral, came down with a flu-like illness: fever and headache for a few days, followed by several weeks of *asthénie* (the elegant equivalent of "feeling crappy"). Since no one died, and the all too real problem of the time was the German occupation–Vichy government, this relatively mild disease was given the Gallic shrug of dismissal. And at any rate, diagnostic virology was still too unformed to do much about it. However, from retrospective review and serology, it is now generally conceded that most of those cases were of West Nile fever.

Twenty years went by without new outbreaks and the Montpelier fevers receded into local medical history. Then in 1962, Montpelier physicians got a booster to their memory bank, but this time the art of virology was ready to meet the diagnostic challenge. Within two years, from 1962 to 1964, there were thirteen cases that had the clinical character of acute West Nile disease—high fever and neurological signs of meningoencephalitis in eight patients, of whom three died. All of those with severe neurological complications were elderly. Alarmed, the French government sent a team of experts to Montpelier.

One of the entomologists was Jean Mouchet, a man so experienced he almost thinks like a mosquito. Mouchet promptly identified the carrier, a local mosquito, *Culex modestus.* And to prove his point, Mouchet and his colleague obligingly came down with West Nile fever. Fortunately, theirs was of the mild variety; but it did show that the onset of symptoms was three to six days after being bitten by an infected mosquito. Also, the virologists were able to isolate the virus and correctly identify its species from their own blood, taken at an early stage of the infection when the virus was still circulating. With this knowledge, French public health authorities began an intense insecticide spraying campaign around Montpelier and the adjacent Mediterranean littoral which ended that West Nile story—until 1965, when it killed Halima II in the Camargue.

The Camargue is the marshy, grassy region in the arms of the Rhone River delta, south of where it splits at Arles. The French consider it a strange, somewhat alien place. The investigators sent to study the West Nile virus there described it as a region of "unquestionable originality" full of "wild, more or less domesticated vertebrates," and a great number of "aggressive arthropods." When my Parisian friends speak of the Camargue, they include the human inhabitants among the "wild, more or less domesticated vertebrates." Lots of semi-wild horses and cattle, lots of hungry mosquitoes to feed on them. In the summer of 1965, one of those horses, a six-month-old colt dignified with the name of Halima II, fell ill and then moribund with what seemed to be a viral inflammation of the brain and its membranous envelope (meiningoencephalitis).

Halima II was euthanized, his West Nile virus isolated and

identified by the Pasteur Institute and the Lyon Veterinary School virologists. To confirm the virus's equine affinities, a donkey, two adult horses, and three colts were inoculated with the virus from Halima II. Over the next six days, all the animals were stricken with a high fever. This progressed during the next few days to signs of central nervous system involvement—somnolence and then paralysis. At autopsy, the brains and spinal cords of these animals showed the inflammatory stigmata of the viral infection. Winter came to the Camargue, cooling the mosquitoes' aggression. The West Nile virus left this "unquestionably original" region, returning to kill horses again in 1999. In the intervening decades, the virus had also moved to another European river delta.

From the Rhone delta to the Danube delta; from Montpelier to Bucharest, is not all that far, as the West Nile virus–infected crow metaphorically flies. Bucharest, once one of the loveliest and noblest cities of central Europe but now sadly shabby after its long ordeal of communist rule, lies near to the Danube delta, where the river debouches into the Black Sea. This great, marshy flood plain is the habitat for a huge population of resident birds, as well as the seasonal haven for birds migrating to and from Africa. It is a bird watcher's paradise, providing the birder can tolerate the blood-thirsty hordes of the marsh-breeding mosquitoes. This is the epidemiological geography of entrenched West Nile virus in the avian population. The people of Bucharest had been spared, the city being beyond the flight range of the marsh mosquitoes. This all changed in 1996.

Hospitals and physicians are accustomed to the ordinary medical miseries—heart attacks, strokes, cancers, and the mayhem of modern life. They react more slowly and stubbornly when confronted with a disease new to their experience, especially when it comes as an epidemic wave. In the summer of 1996, the hospitals of Bucharest were overwhelmed by patients with a new, apparently infectious syndrome. It began with the "flu"—moderate to high fever, sore throat, loss of appetite, and fatigue commenced to "dengue"—rash, muscle and joint pains, and sometimes, respiratory distress. At its end, the epidemic left about 10 percent of the patients with encephalitis/meningitis: stiff neck at the onset, confusion, tremors, convulsions, a coma, and death. The guilty pathogen was not among the usual suspects, and the Bucharest microbiology laboratories did not have the facilities to search for the "exotic." So they called on their friends for a little help.

Globalization was the business of science centuries before it became the business of business. Networking has been science's tradition and strength. If there was ever a need for international cooperation, it was in the 1996 Bucharest outbreak. There was a new urgency; the West Nile fever was no longer only a flu-like childhood illness. It had, somehow, changed into a potentially lethal neurologic disease. It was a new, true epidemic with over 10 percent of the Bucharest inhabitants infected. The first outreach was to the French virologists at the Pasteur Institute who were sent a sample of cerebrospinal fluid from a fatal case. From this, they grew a virus in tissue culture and identified it as the West Nile flavivirus. With the knowledge that it was an arthropod-transmitted infection, an American collaboration began with the entomologists, led by Dr. H. M. Savage, from the Division of Vector-Borne Diseases of the Centers for Disease Control and Prevention in Fort Collins, Colorado.

Not only had the virus seemingly changed during the thirty years since the 1965 mini-outbreak in southern France, but so had virology. New, powerful, probing techniques of molecular genetics had made possible a new approach to epidemiology—a molecular epidemiology. The DNA sequencing methods that convict the rapist, exonerate the death-row inmate, or force the denying father to write child support checks can also be used to trace the lineage of viruses and other microbial pathogens. A West Nile virus from India and a West Nile virus from Kenya are both West Nile viruses, but their genes encode small differences that make them identifiable as to their geographical origin. When the gene that directs construction of the virus isolated from a Bucharest mosquito was sequenced, it revealed it to be of tropical African lineage. This was evidence that the Romanian outbreak was caused by infected, carrier birds migrating from sub-Saharan Africa. But why 1996? Surely infected birds had been flying to the Danube basin for countless centuries, but this seemed to be the first time that it resulted in a human epidemic of a severe form of the disease.

The first logical, fashionable, almost reflexive explanation given was global warming. More heat and more rain equals more mosquitoes breeding and more mosquitoes carrying virus. To prove the global warming hypothesis, meteorologic data—monthly temperatures and precipitation—of the epidemic year was compared to that for the preceding eighteen years (interestingly, these data were archived not in Romania but at the National Climatic Data Center in Asheville, North Carolina). But the data didn't support the theory; the 1996 transmission season (May through October) was much drier than the rainfall averages for the nonepidemic years. Average daily temperatures didn't vary either except that it was slightly warmer in June and July of 1996.

If not gross climatic abnormalities, what then? Environmental factors are usually the next suspects, and this may be closer to the mark in Bucharest. For their workers' paradise, the Romanian government had built many high-rise apartment buildings, concrete blockhouses in the style of the Lubyanka Prison school of architecture. Single-family homes with pit toilets, chicken houses, and sometimes, stables remained in the high-rise neighborhood. Serving the new apartment houses was a very old, defective sewer system. Savage and his coauthors described the plumbing of Bucharest in these words: ". . . both water and sewage pipes had deteriorated and the basements were partially flooded with a mixture of drinking water and raw sewage." Dirty water is the habitat of a major West Nile virus mosquito vector, *Culex pipiens,* whose larvae thrive in this rich organic muck (and also, of course, in the pit toilets). In entranceways, halls, and in the apartments themselves, thousands of *Culex pipiens* were laying in wait for a warm-blooded Romanian.

Linking the hypothetical environmental factor to a hypothetical weather factor, epidemiologists produced a hypothetical cause for the Bucharest epidemic of 1996. It was believed that the driest summer in eighteen years was actually a major contributory factor. The lack of rain didn't flush the sewer system, as it normally does in the summer, sweeping out many of the mosquito larvae. The residual small, stagnant pools were even a more favorable breeding habitat for the *Culex pipiens* larvae. Then, that slight elevation of summer temperature was critical in the enhanced survival and transmission of the virus in the mosquito. The lesson of Bucharest is that it is not big, dramatic changes that may cause outbreaks of vector-borne diseases; a package of small, subtle perturbations can have the same epidemiological effect.

Chapter

2

West Nile-on-the-Hudson: The Virus, 1999 and Beyond

From the flushless sewers of Romania, we come to the West Nile virus's 1999 attack on Flushing, New York, its debut in the New World. In the summer of 1999, birds and humans in the New York area, particularly Queens, sickened and some died of a neurological disease resembling meningoencephalitis. The storyline for popular consumption would have us believe that with great speed the public health laboratories, led by the Centers for Disease Control and Prevention (CDC) swiftly applied the most refined, sensitive, sophisticated methodologies which quickly identified the causative pathogen in both birds and humans as the West Nile virus, an organism not previously known to be present anywhere in the New World. This scenario concludes with the scenes of a rapidly mobilized epidemiological surveillance followed by spraying with nontoxic insecticides, strategically applied to control the outbreak.

A more accurate documentary would be a cautionary tale of the West Nile virus outbreak seen as a shambles—a chaotic confusion that, in retrospect, is truly frightening. If the West Nile virus is a curtainraiser to the arrival of a truly nasty alien pathogen, like the Ebola virus, then we are in big trouble if we are to depend on governmental services to protect us—a conclusion also arrived at by the thorough inquiry of Congress's "detective agency," the General Accounting Office of Congress, and the Senate investigation requested by Joe Lieberman of Connecticut.

In the beginning, there were two epidemics of the same epidemic, one of birds, the other of humans. The initial epicenter of both was in Queens, the borough of two World's Fairs, the Shea Stadium home of the Mets, the U.S. Open Tennis Stadium, and LaGuardia Airport. Queens can also boast hundreds of acres of bird-rich marshes.

The first sign of the virus occurred in early June when concerned residents of Queens brought dead and dying birds to the Bayside veterinary clinic where a veterinarian noted symptoms of a neurological affliction. Over the next month, other birds were found dead in many New York City parks. The supposition was that they had been poisoned either maliciously, or worse still, by some environmental mishap. A veterinarian at the Wildlife Conservation Center (a unit of the Wildlife Conservation Society based in the Bronx Zoo and concerned with wildlife welfare in and out of zoos throughout the world) in the zoo at Flushing Meadows, Queens, had been treating sick wild birds. He sent specimens to the New York State Department of Environmental Conservation, whose laboratory was expert in toxic analyses, because the cause of death was surmised to be poison rather than pathogen. Their investigators found no evidence of poison in the dead birds, and no common thread to their deaths. Most of the dead birds were unloved, but highly susceptible, crows. From New York to Buffalo, people brought bagfulls of dead birds to their local health authorities, and by the end of summer at least 17,000 crows had died, maybe half of New York State's crow population.

August 1999 was also the cruelest month for humans in Queens. On August 12, the first of many patients was admitted to the relatively small Flushing Hospital Medical Center. This sixty-year-old "well-tanned" man had been weak, feverish, and nauseous for three days. The next day, he became weaker, confused, lost his deep-tendon reflexes, and lost his ability to urinate. He survived and is now making a slow recovery but remains weak and has episodes of memory loss. On August 15, an eighty-year-old man (also described as being "well-tanned") was brought to the hospital by his wife who found him "unresponsive" after a week's bout of weakness, fever, nausea, and diarrhea. In the hospital, he rapidly went downhill and was declared brain dead three days later. Three weeks later, the life support systems were turned off and he died. On August 18, a seventy-five-year-old man with similar neurological symptoms entered Flushing Hospital. His condition progressively worsened and he died.

These cases, in a small community hospital, might well have gone unnoticed or been viewed as random events—old people having a bad time in a very hot summer—if it were not for a very alert staff physician. Dr. Deborah Asnis, an infectious disease specialist at Flushing Hospital, noted the cluster of patients with similar symptoms admitted within a short period of time. The fever, the weakness, the neurological symptoms pointed to a microbial etiology. The blood and cerebrospinal fluid of all the patients were sterile, the specimens did not grow bacteria or fungi in culture. This left a viral cause, but what virus? Her hospital laboratory lacked the technology to make the serological diagnosis for an unknown virus. On August 20, Dr. Asnis phoned the Bureau of Communicable Diseases of the New York City Health Department and conveyed her suspicions and anxiety that Queens was experiencing an outbreak of an unknown disease which could prove fatal in old people. Help! On August 23, a fifth patient was admitted, and on that day, an epidemiologist from the city health department came to the hospital and advised that cerebrospinal fluid specimens, on dry ice, be sent to the health department's laboratory. The suggestion was also made that they should rule out botulism food poisoning by trying botulism antitoxin, since nothing else worked. Flushing Hospital did not have any dry ice and had trouble getting it.

On August 27, two weeks and a day since the first patient's hospitalization, the director of the city's bureau of communicable disease decided to go to the ultimate referral source, the Centers for Disease Control and Prevention (CDC) in Atlanta, Georgia. The CDC's Foodborne and Diarrheal Disease Program, which deals with botulism outbreaks, was contacted and the Flushing patients described. The CDC people opined that it didn't sound like botulism but it sure was strange. The director of the city's bureau of communicable disease now thought it time to do field work, and the next day, Saturday, August 28, he traveled to the Flushing Hospital where he found the newly admitted sixth patient with symptoms suggestive of a viral encephalitis. Urgent calls to other hospitals in Brooklyn and Queens turned up three more suspected cases diagnosed as presumptive severe neurological disease of viral origin. An outbreak of a viral, communicable disease now seemed an ominous threat to New York City; the identification of the causative pathogen became absolutely essential to formulate and implement public health strategies. Again, the city went to the prime referee, the CDC, but this time to their Division of Vector-Borne Infectious Diseases, Viral and Rickettsial Diseases, and the Bioterrorism Preparedness and Response Project.

The CDC is a large, rambling, multipurpose organization. Their Chamblee facility in suburban Atlanta is a center for epidemiological surveillance. It is also a major center where basic research is carried out at the molecular and population levels. In malaria, for example, Dr. John Barnwell searches for the ligands binding the malaria parasite to the red blood cell; Dr. William Collins raises malaria-transmitting mosquitoes and uses them to test candidate antimalaria vaccines in monkeys; and Dr. Richard Steketee studies the mechanisms that are responsible for severe malaria in pregnant women, especially during their first pregnancy. Atlanta CDC can dispatch expert epidemiologists to disease outbreaks not only in the United States but throughout the world. There is a zoonosis division where Dr. Peter Schantz, a veterinarian researcher, is trying to figure out how and why a sandfly-transmitted (*Phlebotomus* species) parasitic protozoan of southern Europe, *Leishmania infantum*, that is potentially devastating to humans and dogs, has come to infect foxhound hunt packs of the eastern United States. In some ways, this is even more mysterious than the arrival of the West Nile. The CDC advises travelers on health risks in countries throughout the world. Should a traveler return with some exotic disease for which the therapy is not stocked by your friendly pharmacist, then the CDC supplies it, from their repository, to the physician. Thus the Atlanta CDC is many things to many men, women, and children, but when it comes to knowledge about vector-borne viruses then we must journey to their Fort Collins facility in Colorado.

The modest building sits on land that appears to be a cow pasture. In the near distance, the Rockies fill the skyline, and a short drive in the right season brings you to the sound of the bugling, lusty elks. The outskirts of Fort Collins seem an incongruous setting for the study of communicable and exotic tropical, arthropod-transmitted diseases, everything from plague to dengue fever. There is however, a historical reason why the nation's first line of defense against domestic and foreign insect-transmitted germs is in the Rocky Mountains. At one time, an important tick-transmitted disease, Rocky Mountain Spotted Fever, was common in the American West. It is caused by an intracellular bacteria-like rickettsia germ, *Rickettsia rickettsi,* with reservoir hosts in many kinds of wild animals. Before the advent of antibiotics, the death rate from Rocky Mountain Spotted Fever was anywhere from 20 percent to 50 percent. Two regional research laboratories were established to study the disease, one at Hamilton, Montana, the other at Fort Collins. The broad-spectrum antibiotics, chloramphenicol and tetracyclines were so effective that they more or less made the laboratories unnecessary. Hamilton was shutdown.* The Fort Collins facility withstood repeated closure threats. In 1994, it got a more or less permanent lease on life when the Clinton

* A possibly apocryphal story has it that the Hamilton laboratory gave us the Waring blender. My friend Dr. Paul Reiter, of the CDC's Puerto Rico unit and a mutual trivia lover, told me this story. The Hamilton lab invented an apparatus to macerate safely infected tissues. It was an upside-down wood-working router enclosed in a large closed jar—a blender. Paul said that Fred Waring, the band leader, somehow saw this machine, improved on it, patented it, and by so doing, enriched and immortalized himself.

administration discovered that Mr. Hussein of Iraq possessed a tanker-full of botulism toxin, and God knows what other deadly biologicals. Virtually any tin-pot tyrant with a germ warfare laboratory could wipe out the United States of America. Prodded by the administration, Congress asked the CDC to develop a national bioterrorism defense strategy. The expertise for this was already at Fort Collins. That facility functions under the forceful direction of entomologist-virologist-parasitologist, onetime cowboy-actor, and formerly a University of Hawaii colleague of mine, Dr. Duane Gubler. Today, it is the major resource for field and laboratory investigations of insect-transmitted diseases and bioterroroism. It is also a major diagnostic reference center for those diseases. In that role, the CDC of Fort Collins first encountered the West Nile virus.

On August 29, an epidemiology investigation specialist from Fort Collins came to Flushing where he was met by representatives of the New York State Health Department. When a new infectious disease comes to town, epidemiologists scramble. They review the records. Then begins the shoe-leather, sometimes door-to-door inquiries examining the environment where the cases occurred. Who first had the infection (the index case)? What was the patient's history? Who else has the infection (its prevalence)? How many new cases occur over a given period of time (its incidence)? To these time-honored methods, modern genetic-molecular techniques have added powerful investigational advantages.

During a neighborhood walkabout, sometime during the previous week, the state's epidemiologist noticed mosquito breeding sites around the homes of patients hospitalized with "Flushing Disease." This gave a hint that the causative pathogen was a mosquito-transmitted virus, a flavivirus such as St. Louis encephalitis and Eastern Equine encephalitis viruses. The New York State Health Department laboratory acting on this clue carried out serological tests on the patients' blood samples for a flavivirus.* They also tried to recover viral DNA from the brain of a dead human victim and to amplify it by a polymerase chain reaction (PCR) so that it could be sequenced to yield the virus's identifying signature. The PCR failed, but the serology was positive, and among the flavivirus test antigens available in the state lab, the strongest reaction was to the St. Louis encephalitis virus. This indirect diagnosis fit the cases—sort of. St. Louis virus is generally mild in the young but severe, sometimes fatal, in the elderly and the immune-compromised. It produces signs, symptoms, and pathology not unlike those in the Flushing cases. It is mosquito-transmitted by a species of mosquito known to be present in Flushing. Outbreaks of St. Louis encephalitis are not common east of the Mississippi but they have occurred in places like Philadelphia and New Jersey. The St. Louis encephalitis virus is a zoonosis with birds as reservoirs although it is not harmful to the infected birds.

When the visiting CDC epidemiologist conferred with the state health department, he was told on August 31 that strong evidence pointed to St. Louis encephalitis virus. Samples were rushed to Fort Collins that day, where they were processed for flaviviruses by the relatively rapid enzyme-linked immunoassay (ELISA). Two days later, September 2, the CDC confirmed that their screening tests also pointed to St. Louis. However, in that screening, the CDC, despite

* Serology, most simply, is the visual measurement of the combination of antigen and the antibody elicited to it.

being the repository for a comprehensive library of flaviviral test antigens and despite its recently adopted slogan, "Expect the unexpected," did not include West Nile virus in the serologies. The possibility of so alien a virus invading Flushing, New York seemed to be dismissed. That same day, September 2, Flushing Hospital admitted its eighth gravely ill, elderly patient with the now presumptive diagnosis of St. Louis viral encephalitis. Meanwhile, the crows kept dying, but neither the CDC nor the state health department associated dying birds with dying humans. Besides, St. Louis encephalitis virus doesn't kill birds.

Finally, it would be the birds and their human advocates, particularly the dogged, nagging persistence of the Bronx Zoo's chief veterinary pathologist that led to the correct viral identification. All those thousands of dead crows were a concern, possibly as sentinels for an environmental poison, and during July and August 1999, they were being autopsied and their tissue analyzed for toxins or poisons by the New York State Department of Environmental Conservation. This was not shrouded in secret; the bird test findings made the newspapers and an excellent infectious disease bulletin board, ProMed (www.ProMEDmail.org), which continues to monitor the West Nile virus as well as other emerging infectious diseases. Concern and interest dramatically intensified when birds began to die at the Bronx Zoo. The problem assumed a new dimension. First to go was a Chilean flamingo on August 10. During the next six weeks, twenty-seven zoo birds either died or were moribund and had to be euthanized. The list of the feathered dead was: 5 Chilean flamingoes, a black-crowned night heron, 3 guanay cormorants, 2 laughing gulls, 2 bronze-winged ducks, 1 mallard duck, 2 Himalayan Imperyan pheasants, a Blyth's tragopan, a bald eagle named Clementine, a snowy owl, a black-billed magpie, and a fish crow. Five dead common crows were also found on the zoo grounds.

During the second week of August, when the New York State Health laboratory was processing the human material from Flushing, the New York State Department of Environmental Conservation pathologist was processing material from the dead birds at the Bronx Zoo. No viral cause for either the zoo's birds or those that died in nature was detected, and it was concluded that there were a variety of causes "with no common thread." During the Labor Day 1999 weekend, several more birds at the Bronx Zoo died, and the stumped Environmental Conservation pathologist sent tissue samples to a federal wildlife laboratory, the National Wildlife Center, which for reasons best known to the federals, is a unit of the United States Geological Service (rocks and animals?). In a covering note, accompanying the tissue and blood samples, the New York pathologist commented that the city was experiencing an outbreak of St. Louis encephalitis. Did the Geological Service experts think there might be a connection to what was happening to the New York region's birds? This seemed to be the first instance that someone thought to associate the outbreaks in the two species of animals—birds and humans. Then, to cover all pathology bets, the New York Environmental Conservation laboratory sent a duplicate set of the bird samples to the United States Department of Agriculture's Veterinary National Services Laboratories in Ames, Iowa.

During that Labor Day weekend, Tracy McNamara was in a stew at the zoo. Another Chilean flamingo under her care had died. Its fixed and stained tissue preparations revealed, under the microscope, lesions in the brain, spleen, and heart, typical of an acute viral infection. Suspecting a human–bird connection, she called the CDC in Atlanta, requesting laboratory diagnostic assistance. The CDC in Atlanta told her to get in touch with the CDC in Fort Collins. She called Fort Collins and spoke to the head of the Epidemiology and Ecology section and told him of her suspicions and need of assistance. He replied, in effect, "Lady, we don't do zoos; go try the USDA lab in Ames."

Dr. McNamara remains to this day one furious vet. She is unforgiving of what she views as the CDC's obtuse unresponsiveness and inefficiency. She maintains that the CDC would do no better if confronted with a truly dangerous zoonotic. She also accuses the CDC of causing some $100,000 loss in visitor revenues to the Bronx Zoo. When the West Nile virus dust finally settled, the CDC began covering their *okole* (to use the Hawaiian idiom) by suggesting that the whole epidemic began from an infected bird at the Bronx Zoo.That's nonsense, claims Dr. McNamara, but its mischief kept people from visiting this alleged primary source of a strange and killing disease.

Rejected by the CDC, Dr. McNamara looked elsewhere for help, and on Thursday, September 9, she shipped tissue and blood samples of a Guanay cormorant, Clementine the bald eagle, a Chilean flamingo, and a pheasant (tragoplan) to the USDA's Ames, Iowa, veterinary laboratory. But she also had a human problem; one of her laboratory technicians had been accidentally stuck with a syringe needle while withdrawing blood from a stricken bird. McNamara was concerned that the employee may have become infected with the X pathogen. This was clearly within the province of the CDC, so she shipped a serum specimen to the CDC at Fort Collins—and for good measure included a serum sample from a dead Chilean flamingo. Prodded from several directions, the CDC began to respond, and on Saturday, September 11, their vertebrate ecologist arrived in New York. But his brief was limited; he tested only live birds.

Meanwhile, at the "human" laboratory at the New York State Department of Health, doubts rose about the St. Louis encephalitis virus. Their tests by PCR did not confirm St. Louis as the causative flavivirus and they consulted Dr. Ian Lipkin, director of the Emerging Diseases Laboratory of the University of California at Irvine, an expert at identifying virus species by genome sequencing. The Health Department laboratory sent Dr. Lipkin brain tissue from a patient who had died of the disease.

On September 14, the U.S. Department of Agriculture's Veterinary Services Laboratory made the first virus "landfall," an isolate from a dead Bronx Zoo bird. They called Dr. McNamara and told her the good news that they had the virus and the bad news that, aside from the broad diagnosis of flavivirus, they hadn't a clue as to its species. Their electron microscope images showed only that it was a "flavi/toga-like virus, 40 nanometers in diameter," but the serological tests for St. Louis encephalitis, Eastern Equine encephalitis, Western Equine encephalitis, and Venzuelan Equine encephalitis viruses were all negative. It was a mystery. Maybe a new strain of St. Louis that could kill birds? With this information, everyone began talking with one another. Like born-again virologists, the CDC at Fort Collins became motivated and requested the bird virus isolate be sent to them. The Ames lab complied with the request . . . a week later; there seemed to be no great rush. On September 21, the isolate arrived in Fort Collins where the CDC began their analytical work.

Dr. McNamara remained dissatisfied with the slow pace of progress. Her cherished birds were dying and she had lost confidence in the CDC. So, on that busy September 14, she contacted the veterinary microbiologists at another expert facility, the U.S. Army Medical Research Institute of Infectious Diseases (USAMRIID) at Fort Detrick, Maryland, an astute move, as there is probably no better facility to unmask the unexpected. In the bad, old unregenerate days, Fort Detrick was the military's center for developing biological weapons, a.k.a. germ warfare. Several treaties later, it has become unthinkable that America would engage in germ warfare or the research that produces its weaponry. But evil still stalks the world, and it is obviously necessary to maintain a defensive research capability. For that defensive research, it is necessary to maintain and develop those germs and toxins that could be used against Americans and their troops.

Now the critical elements for discovery were in place. The bird and human samples, as well as the virus isolates, were in the hands of experts working in laboratories equipped with all the toys and gadgets necessary to carry out the techniques for viral identification. There was now recognition that this was a zoonosis, that the same virus was probably infecting both birds and humans. September 23 was WNV day; the West Nile virus was unmasked by the CDC's Fort Collins lab. Using a bird virus isolate sent to them by the USDA, they sequenced the RNA that had been amplified by the polymerase chain reaction (PCR). It was most like a West Nile virus; it was not a St. Louis encephalitis virus. Concurrently, Dr. Lipkin in California, who had been working on the virus from a human case, also completed his sequencing analysis and compared it to other flavivirus sequences that were in the national genomic data bank. It corresponded to a West

Nile virus. Confronted with the unexpected finding, the CDC retested the human viral material by genomic and protein envelope-sequencing and confirmed that the West Nile virus was a true bill. Teleconferencing informed all parties of the very strange, unexpected immigrant virus that had come from the Old World to the New World's northeastern United States, where it had killed birds, humans, and horses (horses were particularly susceptible; seventeen died on Long Island alone). There was a consensus that there should be a news release. The CDC was not quite ready to eat crow and favored saying only that the West Nile virus was responsible for the bird mortality. However, the true story could not be contained, and over the next two days, the media broke the news that the virus had been found not only in birds but also in humans. By September 25, the world came to know the bizarre truth.

"Bizarre" it may have been, but so was the Black Death when it came to fifteenth-century Europe. In comparison, we are fortunate that it was the relatively benign West Nile virus that was America's trial run for our readiness to diagnose and contain the unexpected pathogen. How then should we grade our medical watchdogs? For the report card, let me synoptically run the New York West Nile story by you again, and you be the judge. There is retrospective evidence that the first bird death occurred on or about May 21 and the first clinically significant human case about eight weeks later. About two weeks later, an alert hospital physician and veterinary pathologist independently called attention to what were considered separate outbreaks in birds and humans. Conclusive identification of the West Nile virus was made during the third week of September, *four months from the death of the first bird*, and one and a half months from the first professional observation of what was happening to birds and people. The final conclusion involved the New York City Department of Health laboratory and field personnel, the New York State Department of Health laboratory and field personnel, the New York State Department of Environmental Conservation laboratory and field personnel, the Connecticut Department of Health field personnel, the Connecticut Agricultural Experimental Station laboratory, the Centers for Disease Control's Foodborne and Diarrheal Disease Program personnel, the Division of Viral and Rickettsial Diseases personnel, the Division of Vector-Borne Infections laboratory and field personnel, the Bioterrorism Preparedness and Response Project personnel, the United States Department of Agriculture's National Veterinary Services Laboratories, the United States Geological Survey's National Wildlife Center laboratory, the United States Army Medical Research Institute of Infectious Diseases laboratory, and the Emerging Diseases laboratory of the University of California at Irvine. I haven't included Dr. McNamara of the Bronx Zoo or Dr. Asnis of Flushing Hospital in this list because I consider them to be "prime movers" rather than "processors." But is this any way to run a business that deals with protecting the public's health?

Great questions remained: where had the West Nile come from and how did it get to Flushing? One version of an explanation is the story I heard at the breakfast table in Lausanne, Switzerland, during a recent malaria meeting. The raconteur was Dr. Duane Gubler, chief of the Vector-Borne Diseases laboratory of the Centers for Disease Control, who with his colleagues were very much involved in assessing the

New York outbreak and who continue to monitor the virus in the United States.

The Flushing outbreak occurred in the late summer of 1999. By December, virologists had isolated the virus from the brains of dead human and avian victims and had analyzed its genetic-chemical constitution in order to determine its relationship to other strains from known endemic regions—to see where it originated. A team led by Dr. X. Y. Jia of the University of California at Irvine cloned a virus from the brain of a dead human and, by an elegant technique known as reverse-transcription polymerase chain reaction, showed that the genome had a lineage nearly identical to West Nile viruses from the Middle East. Duane Gubler's group got their virus from the brain of a Chilean flamingo that had died at a New York zoo. They took a different analytical approach, stripping off the covering envelope of the virus, a glycoprotein, and sequencing its amino acids. This protocol revealed that the New York virus not only came from the Middle East but it was the same virus that had killed a goose in Israel in 1998. And therein hangs Duane's tale, told to me over breakfast at the Lausanne Palace Hotel and Spa on a bright May morning of 2000.

Israel is the crossroads for many birds moving seasonally from Africa to Europe. There is even a Migratory Bird Watching Festival in Eliat, on the Red Sea. With all those African birds touring Israel, it is not surprising that some of their West Nile virus spills over to infect local mosquitoes. These in turn infect local birds—and local Israelis. Hundreds of cases of West Nile fever have been reported in Israel. As if Israel didn't have enough troubles, the country has experienced repeated epidemics of West Nile fever, often around the times of conflict, including a new mini-epidemic during intifada II of 2000. In 1989, the Israeli Army medical corps reported an outbreak in the Negev, a place where thousands of African raptors congregate during migration—a wondrous sight. From 1997 through 1998, an avian outbreak killed hundreds of geese. But these were not your run-of-the-mill honking, hissing Jewish geese; they were "industrial" geese, bred and fed to produce the fatty livers that are the substance of pâté fois gras. Not well known as a commodity of international trade, the Israeli pâté fois gras is exported to France (where it presumably is sold as the classical Alsatian product).

In late 1997 into early 1998, Mad Cow disease was at the height of its notoriety; the French had embargoed the import of British beef. The Israelis, as the story goes, were alarmed. Despite all biological irrationality, the French might view the anserine West Nile in a similar light—as a Mad Goose disease—and close this profitable trade. So, while the Israelis didn't exactly hide the fact that West Nile virus had decimated their goose flocks, they certainly didn't publicize it either.

During this time, a female French veterinarian who was a visiting researcher in Israel got hold of an infected goose brain which she "illegally" sent back to France (it is not certain whether the Israelis would have banned its export or the French banned its import as biologically dangerous). In the French laboratory, the virus was isolated and its genome and glycoprotein envelope characterized.

In New York, birds and humans had been stricken with the West Nile virus. In more innocent times, it would be considered an unusual occurrence, but probably no great effort would be made to pursue its exact origins. Now however, the threat of bioterrorism concerns virus hunters when Americans are struck with a seemingly new infectious agent. Was the pathogen deliberately introduced by an enemy?* Suspiciously, the glycoprotein sequence of the West Nile-on-the Hudson did not match up with that from any other identified strains of the virus. So, where did it come from?

This American outbreak of an African-based, mosquito-transmitted virus demonstrated once again that it is a small world for the ARBO viruses; that wherever there are competent mosquitoes there is the potential for transmission. But it is also a small world for the ARBO virologists. It is not a big calling, and they all know each other pretty well. And now, the Internet has knitted them, as it has for so many scientists with mutual interest, into a large, international family whose informal discourse flows as freely on the computer screen as it would over the dinner table. When CDC investigators began their informal international inquiry with "Has anyone seen this virus?" they got a response from the French who said that the Flushing West Nile's glycoprotein sequence matched the virus obtained from the brain of the Israeli goose.

This intriguing bit of information still hasn't solved the mystery of how, what, or who brought the West Nile virus to the American shores. An infected immigrant/foreign visitor/American tourist?† An infected African bird blown by a storm to Flushing? Infected mosquitoes carried by cargo

* The October 1999 *New Yorker* magazine had a "provocative" article asserting that the West Nile virus was introduced by foreign terrorists; but surely, even the likes of Saddam Hussein and Osama bin Laden could do better than targeting eighty-year-olds in Flushing, New York.

† This happens. In 1984, a sixty-eight-year-old French woman returned home after a month's stay in Israel. She came down with a severe infection of the central nervous system that at first resembled polio. Serological tests proved she had West Nile viral meningomyelocephalitis.

ship from Africa or some other endemic focus to the docks of New York? Certainly, the epidemiological concatenation of the virus's variety of avian and mammalian hosts and the numerous variety of mosquito and other blood-sucking insects capable of transmitting the infection make the West Nile virus a potential pathogen. It is just amazing that it didn't appear in the United States or elsewhere in the Western Hemisphere before 1999. West Nile fever can be a nasty disease, but for the most part, it is lethal only among the aged and infirm. However, if it mutates or is made to mutate by an evil biogeneticist bioterrorist into a highly pathogenic form, then we are in for some big trouble.

With curiosity ostensibly satisfied as to where New York's West Nile virus came from, if not how it got there, we can attend to the more practical issue of what can be done about it. Should anything be done about it at all? Has there been panic without reason? For the human populations at risk, be they in Nairobi or New York, there is no available protective vaccine, and there may very well never be. The virus's relatively low pathogenicity for humans has restrained industry, military, and institutions such as the National Institutes of Health from seriously undertaking the long, arduous, dubiously possible or profitable effort to produce a safe, effective vaccine.* Nor is there any specific chemotherapy. A serological survey showed an estimated 1.4 percent infection rate among the residents of Queens, or 1,903 infections when extrapolated to the entire exposed population number. In 1999, there were only sixty-two people whose disease was

* It may be a telling insight into commercial biomedical economics that the first available West Nile vaccine is for horses and not for humans. In 2001, the Fort Dodge Animal Health Laboratories, Inc., Iowa, began selling a killed-virus vaccine to protect horses. Immunization requires two injections, six weeks apart.

severe enough to require hospitalization and of these, seven died. The fatal cases were almost exclusively in the very old, with an average age of eighty years. The vast majority of the infected (serologically positive) young and middle-aged experienced no discernible illness or symptoms more severe than the "take two aspirin" variety. We might well ask what should be done to combat a vector-borne microbe that causes a 1.4 percent infection rate of which only 3.2 percent are symptomatic, with a fatality rate of 0.3 percent (case fatality rate of 11.2 percent). Did the mayor of New York and his health advisors unnecessarily push the panic button on the spray can to release a cloud of insecticide?

The anxious public needed visible, dramatic proof that the municipal government was acting to contain the epidemic. For this, there was only one option. A flavivirus, even the erroneously accused St. Louis encephalitis virus is transmitted by mosquitoes. The major mosquito vector of both St. Louis encephalitis virus and West Nile virus is *Culex pipiens*. The *Culex pipiens* female lays her eggs in stagnant, polluted pools of water where the larvae emerge, swimmingly thrive, pupate, and emerge as adults. Sewage seepages, tin cans, discarded tires, derelict buildings, unnoticed collections of water in cellars are all typical *pipiens* breeding habitats. The summer of '99 was hot and dry. Many small collections of water remained unflushed by heavy rains. Also, Mme. *pipiens* is a vampire who sleeps by day and flies and feeds at night, sucking blood from humans, horses, birds—any warm blooded creature.

Within forty-eight hours of the St. Louis encephalitis viral etiology announcement, New York City had contracted with commercial companies (unlike Connecticut and New Jersey, New York had no antimosquito operational facilities) to put

Queens and part of the Bronx under a cloud of insecticide. On September 3, an armada of planes, helicopters, and trucks began intensive midnight spraying with malathion. Spraying continued nightly for twenty days. Mayor Rudolph Giuliani advised those coming under the cloud to stay indoors at night when operations were under way. He also reassured the citizenry that they should have no fear; malathion was a safe, harmless insecticide. But he cautioned of "a slight chance that if you were to just breathe it in, you could get sick." Aside from splitting the infinitive, was he entirely candid?

When that eminently safe and effective mosquito killer, the chlorinated hydrocarbon DDT, was itself killed off in the 1960s, desperate mosquito-abatement personnel turned to the organophosphate malathion which had been developed in 1950 by the American Cyanamid Company. The EPA still considers malathion a safe insecticide if stored under proper conditions. Heat and prolonged exposure to sunlight, however, degrades malathion into a compound some fifty times more toxic than the parent molecule. Nevertheless, malathion has been a workhorse insecticide for many years, killing not only mosquitoes but all sorts of agricultural pests such as fruit flies. I write this in my office/library/refuge. About the room, there is that musty smell so characteristic of malathion. It is December, and the orchids have been brought indoors to wait out the winter. They have just been sprayed with a malathion preparation bought at the local WalMart, and I fear no evil or allergy. One must take chances on behalf of one's loved ones. But then again, my wife thinks I am microbiology's Evel Knievel. So malathion is *relatively* safe but not innocuous. The elderly are particularly vulnerable to its toxic effects. Well, getting old is not for sissies, especially in Queens. First, you are at risk of catching a fatal case of West Nile fever, and if you escape the virus, you might become seriously ill from the malathion that is supposed to prevent you from catching it.

Malathion selectively kills arthropods. A mosquito is an arthropod. So is a shrimp and a lobster. In 1971, the Texas estuarine marshes were sprayed with malathion for mosquito control. About 50 percent of the commercial shrimp harvested there died. In 1999, following the New York City malathion spray operations, 10 million pounds of lobster died in the waters off Connecticut and Long Island. The furious lobstermen blamed malathion, but marine biologists found the cause to be microbial pathogens. Other marine biologists maintained that the malathion contaminated the coastal waters, compromising the lobsters' immune system and leaving them vulnerable to opportunistic pathogens. Still other marine biologists countered that the concentration in these waters was too low to have any effect. Homeopathic marine immunology? Go figure.

When New York City's assault on *Culex pipiens* concluded on September 23, 1999, the outbreak had assumed a different character. The West Nile virus was now known to be the etiologic agent in both birds and humans. Therefore, the regional avian deaths in Westchester County, Staten Island, New Jersey, and Connecticut meant that the virus had spread and those communities would also have to carry out antimosquito measures. And there's the rub. St. Louis encephalitis virus and West Nile virus are both transmitted by *Culex pipiens*, but the West Nile virus has an important and efficient additional secondary transmitter, *Aedes vexans*. *Vexans* is a different animal in both habit and temperament. Breeding in the open waters of marshes and rivers, this is not a species like *Culex pipiens*, which can be controlled by emptying tin cans or old tires, mopping up in the cellar, or flushing sewer systems and drainage ditches. *Vexans* is a very aggressive mosquito which has been described as "one of the fiercest day biters and exceedingly abundant; it is truly a vexatious mosquito." Did you note the "day biter"? *Culex pipiens* bites at night, so when St. Louis encephalitis virus was *the* virus, the common wisdom was to stay indoors in the evening as a mosquito-imposed curfew. Now there was a virus with two vectors, one biting during the day, the other biting at night. When then should spray operations be carried out? It is best to coincide spraying at a time when the mosquito is active, but now there were two species of vectors with different behaviors. Against *Culex pipiens*, the late-night spraying was convenient; people were indoors, often asleep, which would mute some objections to malathion. If this didn't halt the outbreak, then a new strategy was needed to deal with the day-flying *Aedes vexans*.

Westchester, Suffolk, and Rockland Counties decided to spray, despite the advent of fall and the end of the mosquito season. However, their health authorities were sensitive to protests leveled against malathion and turned to the insecticides of better odor, the pyrethroids. The national flower of Japan, symbol of its royal family is the chrysanthemum; it kills bugs. The flower's insecticidal principal (it has a knockdown action) is known as pyrethrum. Pyrethrum has been used in household sprays for many years, but its short life and rapid degradation has limited its use for the control of medically important insects.

Pyrethrum has now been synthesized and the molecule fiddled with to give a cheaper, more stable, longer-lasting family of compounds, the pyrethroids. Some repel as well as kill, and currently there is great hope that bed nets dipped in a pyrethroid solution can significantly reduce childhood mortality from malaria in hyperendemic regions such as sub-Saharan Africa. Pyrethroids are, to use the escape word, "relatively" nontoxic; after all, they are applied directly to cats and dogs to kill fleas and ticks. Nevertheless, in some people pyrethroids may cause dizziness, runny nose, numbness, and itchy skin eruptions. Pyrethroids are deadly to fish.

Industry has bestowed wonderfully belligerent names on the two pyrethroid products used in the New York counties. Surmethrin is sold as Anvil and Resmethrin as Scourge, Crossfire, Derringer, and (best of all) Sun-Bugger #4? Can't you just imagine an American malaria team coming to a former British colony, their trucks loaded with forty-gallon drums prominently labeled SUN-BUGGER #4? For those other former British colonies, the counties of New York, Resmethrin was dispersed in an ultra-low-volume fog from trucks and helicopters. Then, with the cold of October, the mosquito season was over; but not before the area bordering 1,500 miles of roads in Westchester had been Scourge-ed.

Almost fifty years ago, participants in a bold Global Eradication of Malaria Program, endemic nations attempted to reduce mosquito numbers to a level that would interrupt transmission. The strategy was to spray, spray, spray for five years. In a few places, it worked; indigenous malaria was eliminated. But in most places there was only temporary relief. Too many mosquitoes survived the insecticide assault, usually by becoming resistant to the DDT and other insecticides. The ambitious global objective failed, spraying stopped, and it was amazing how rapidly mosquito numbers rebounded. From past experience, a few weeks of malathion, Scourge, and Anvil in New York could not be expected to accomplish what a great international effort failed to do. Many Big Apple mosquitoes survived the ad hoc attacks, and being resourceful New Yorkers, when the cold came, they retreated to winter quarters. During the winter of 1999–2000, inspectors found somnolent mosquitoes resting indoors in cellars and other protected places.

The virus was also wintering over in a protected place—within the mosquito. Some hibernating mosquitoes had earlier acquired the virus from a blood-feed on an infected, viremic animal or human. Other mosquitoes had, however, acquired the virus from an infected mother who passed it to her offspring through the egg, a process known as "vertical transmission." During the 1980s and 1990s, laboratory experiments on insectary-reared female mosquitoes inoculated with West Nile virus showed that their eggs became infected. And subsequently, so did the larvae, pupae, and emerging adult. The virus seemed to do no harm to any of these developmental stages. An inventive field study in Uganda confirmed the laboratory findings. Wild male mosquitoes of known vector species were captured, killed, ground up, and the macerate tested for the presence of the West Nile virus. Some macerate pools were found to be positive, which indicated that the vegetarian male could only have become infected from its mother by vertical transmission.

When the warmth of spring returned to the New York area in the year 2000, the mosquito and its companion West Nile virus were ready to go about the business of perpetuating their lives. However, clinical and public health workers were no longer naive, unknowing, or unprepared and they were ready to reengage the adversaries. With the coming of warm weather, the epidemiologists awaited the unfolding of events. Did the spray rounds of 1999 have an impact on continuing transmission in 2000? If there was continuing transmission, had the virus spread? A multimethod strategy was applied to answer these questions. Mosquitoes were captured, their bodies ground up, and the pools of macerate tested for the virus. Then the chicken was pressed into sentinel service. The chicken is susceptible to infection with the West Nile virus which circulates in its blood stream, a viremia, for a few days without causing demonstrable illness. The chicken's immune system, and it has a very good immune system, activates to kill the virus, leaving only the residual trace of specific antibody as a sign that the virus had come (and gone). The tissues and serum of dead, sick, and trapped birds and animals were also processed for virus detection. Doctors and hospitals now aware of West Nile virus were able to consider it in their differential diagnosis in patients with suggestive symptoms.

The first year of the second millennium ended with ample evidence that the West Nile virus had settled in for a long, if not permanent, stay in America.

THE WEST NILE SCORE FOR 2000

19 severe neurological human cases in the New York/ New Jersey area (Brooklyn 2, Manhattan 1, Queens 1, Staten Island 10), and one mild case in Connecticut;
24 dead or euthanized horses in the tri-state area;
1,271 serologically positive dead birds in the eastern seaboard states from Connecticut to North Carolina;
2 cats, 2 racoons, 3 squirrels, 1 chipmunk, and 14 bats. These animals were all serologically positive but not all with severe neurological disease; however, the diag nosis was important because in many respects the symptoms mimicked rabies.

There is no comfort in the West Nile story for 2001. During that year, the virus broke out of the Northeast corridor to invade as far west as Wisconsin, where dead crows signaled its presence, and as far south as Florida where crows, horses, and people were stricken.

THE WEST NILE SCORE TO NOVEMBER 1, 2001
(WHEN I "CLOSED THE BOOKS")

38 humans (Connecticut 5, New York 6, New Jersey 6, Maryland 6, Georgia 1, Florida 10, Pennsylvania 3, Louisiana 1);
159 horses from 13 states;
5,433 dead birds, mostly crows, reported from 25 states and the District of Columbia, ranging as far west as Missouri.

Of the human cases, only one was fatal. This was a seventy-one-year-old African-American woman living in an Atlanta, Georgia, senior citizen's residence who had been there for the entire month before entering the hospital on July 31. She died of swelling of the brain and its covering membranes (meningoencephalitis) twelve days later.

The West Nile virus obviously has come for a long, probably permanent stay in the United States. Within two years of arriving in New York as an immigrant from the Middle East, it has traveled over 1,000 miles east and south. No amount of antimosquito measures seems to be able to halt this inexorable spread.

It may well be that the most anxious have been the horse owners. Where I live in North Carolina, my horse-owning friends have an extraordinary devotion to their animals. They are greatly concerned that equines seem to be as vulnerable to West Nile as crows. Like the reaction of the French to British Mad Cow disease (before the French knew their herds were also affected), Saudi Arabia banned the import of American horses and Mexico embargoed the import of American chickens for fear of their introducing the West Nile virus. It just makes no sense.

Responding to the single human case in Manhattan and the ten in Staten Island, New York City contractors sprayed again. On July 27, 2000, helicopters rose and laid down wide swaths of Anvil in Brooklyn, Staten Island, and Central Park. In response to the spraying, environmentalists and sensitive New Yorkers formed a No-Spray Coalition. The coalition, in turn, formed an alliance with the faculty and students of the Pace Environmental Litigation Clinic of the Pace School of Law who, in turn, sued in federal court to stop all spraying. The court rejected the argument and in doing so sent a message to the states, "Spray if you must; it's constitutional."

When winter comes, can the mosquitoes be far behind? By the end of next summer, there surely will be more dead birds, horses, and a few old humans, victims of the West Nile virus. New American regions will be invaded by the virus. There will again be fogging of malathion, Scourge, and Anvil, with additional potent larvicides added to the breeding waters. Eventually, there may be a vaccine like the one now available for horses. Equestrians have taken care of their own. Eventually, the surviving birds will be immune, even crows, and population numbers will recover. A few, old and immunologically unsound humans will continue to die each summer. In time, if the West Nile virus does not mutate to a more pathogenic type, it will be viewed as a relatively minor medical problem.

The real issue rising from the alien outbreak is the state of our readiness to meet the challenges of the future, deadlier "West Nile viruses." Both the GAO and Senator Lieberman's committee cautioned that both state and national health entities need to be more alert to the unexpected. Closer cooperation and communication between medical and veterinary groups is essential in both the field and the laboratory. There should be more trained people, better laboratories, and more funding. Certainly, our "National Health Defense Police," the CDC, should have profited from the hard lesson of the West Nile virus to expect the unexpected. But are they a sufficient bulwark in case the next unexpected is a truly terrible plague?

Now is the time to reexamine the state of public health in America. There are exceptions, but over the past several decades, teaching and research on infectious diseases has declined at our schools of public health. They have become more concerned with the cosmic issues of public health policy rather than the nitty-gritty of communicable diseases that once was so prominent in their laboratories and classrooms. Many clinically trained infectious disease specialists are devoted to patient care, monitor hospital-related infections, look after the dialysis patients who are susceptible to microbial infections and, frequently, fill in their income with primary care. Some, like Dr. Deborah Asnis of Flushing Hospital, are alert to the unexpected. But we can't rely on chance to always provide us with someone like Dr. Asnis or Dr. McNamara to sound the alarm. Are there sufficient medically trained professionals who are both public health specialists and monitors of what is taking place in the clinical arena? Should the medical profession be augmented to give us the wide protection needed in this age of the unexpected?

For those of us who have sat for one or two hours beyond

appointment time in order to have a five- to ten-minute visit with a busy, too-often impersonal physician, or in order to have a two- to three-month delay in making an appointment with a specialist, doubt must have crossed our minds. Perhaps the free marketers languishing in their doctor's office have wondered if this is a profession where market forces have been restricted. Those who sit on medical school admission committees know very well that there are at least twice as many candidates as those admitted who are intelligent enough and motivated enough to complete the medical school curriculum and make good, caring doctors. Innovative teaching methods can now accommodate double the student intake. This may force some of the doctors in the new, bigger pool to switch from BMWs to Buicks, but there may also be a sufficient number now in the pool to encourage more to become public health practitioners. About twenty years ago, I was a World Health Organization (WHO) short-term consultant to Bangladesh. My WHO liaison was the Czech epidemiologist, Dr. Karel Markvardt, with whom I mostly liaised in the evening on the tennis courts of Dahka's Intercontinental Hotel. Karel, an MD, told me over a post-set beer that the medical education system in Czechoslovakia was then in two streams, one for future clinicians, the other for future public health specialists, like himself, who had received the training that would also allow clinical practice. Such a two-stream system would be a radical change in American medical education and practice, but perhaps the time has come to contemplate radical changes in the ways we treat illness and preserve health in these times of microbial globalization. If we are very lucky, no great emerging tropical killer diseases will strike us in our temperate zone sanctuaries. But don't bet your life on it.

Chapter

3

The DDT Jitters

Saint Rachel, you will be happy to know that the ospreys are flourishing. But so are those other flying creatures, the anopheline mosquitoes whose unregulated breeding now bring 300 million cases of malaria each year, causing 3 million deaths, mostly of children and pregnant women. After thirty years, we are again engaged in the DDT debates with fundamentalist partisans trying to load the seesaw—environmental pollution on one end, human health needs on the other. Passions have always run high. Allen Ginsberg, the poet, was once our guest when we lived in Hawaii. There was a convivial evening at a local restaurant during which we had a heated discussion on osprey/eagle death by DDT (Ginsberg) and children's death by the mosquito-malaria partnership (Desowitz). Some twenty years later, I still vividly remember the tubby, bearded poet stimulated by a surfeit of poi and wine, angered by the suggestion of an insecticide assault on nature, doing a raptor dance down King Street and shouting "I am the bald eagle! I am the bald eagle!"

Consider the inventory of diseases borne by those blood-sucking mosquitoes, flies (sand and tsetse), gnats, bugs, ticks, lice, and fleas. Thailand alone had 50,000 cases of dengue "breakbone" fever last year. Some of the children went into dengue shock and died. Dengue's supreme carrier, the Asian tiger mosquito *Aedes albopictus*, which arrived in California in 2001, came as immigrants in shipments of "lucky bamboo" (a dracaena, not a bamboo) sent from Thailand, Malaysia, and Indonesia in boxes containing two to three inches of standing water. Yellow fever is on the rise again, killing a few humans and many monkeys in South America and Africa. There are the brain-loving viruses of Eastern equine, Western equine, and St. Louis encephalitides. We squeamish males have apprehensions of the scrotum, so enlarged by the mosquito-transmitted filarial worm that for mobility it may need to be put in a wheelbarrow. And, of course, there is the West Nile virus. That's only a few of the mosquito-transmitted infectious diseases.

Moving on in medical entomology; an assassin bug delivers a protozoan parasite that is the main cause of heart failure throughout a vast region of Central and South America. The tsetse fly, the bringer of sleeping sickness, continues to rule tropical Africa. There are the bubonic plague and its flea, typhus and its louse, the Lyme spirillum and Rocky Mountain spotted fever rickettsia and their ticks. There are the many forms of leishmania and their sandflies destroying spleen, liver, and skin. A particularly nasty form, *Leishmania infantum*, in the children of the Mediterranean littoral, has mysteriously come to infect the foxhound packs in Virginia. This is only a small portion of viral, bacterial, and parasitic arthropod-borne infections. I could go on for several more pages, but this list should be sufficiently convincing that the control of the spineless blood-suckers is essential to human welfare. Nothing has ever equaled DDT for that service, and no essential public health measure has been so irrationally denied.

To begin at the beginning. It is 1940; World War II is raging; Paris has fallen; London is going up in flames; the civilized world is in peril of being overwhelmed by German barbarians, and the Swiss are battling clothes moths. During that year, Paul Müller, a chemist with the Swiss chemical-pharmaceutical firm Geigy, synthesized a chlorinated hydrocarbon compound, dichlorodiphenyl trichloroethylene, which came to be known as DDT. In the Swiss patent, DDT was claimed to have remarkable and unique insecticidal activity. It was the first compound that would kill insects for as long as six months after it was applied. It was a contact killer, requiring no poisonous cloud. It was odorless, and animal tests indicated that it would be nontoxic to humans. This was the ideal product to combat clothes moths.

It was also obvious that DDT could be used against mosquitoes and other arthropods of medical importance. By 1942, the tides of war began to turn against Germany, and the astute Swiss began passing "favors" to the American embassy in Bern. One such favor was the "secret" of DDT, which was given to our military attaché (another story has it that Geigy sent a sample and recipe to their New York office). DDT production in the United States was begun as a pilot project and shortly thereafter used to control louse-borne typhus in the hordes of refugees and concentration camp victims. Those old enough will remember the time when our visual news came as one of the movie "short features." The Movietone/Pathé newsreels showed American soldiers blowing DDT inside the clothing of the war-torn refugees of Italy and Germany. The war came to an end, the United Nations was created, and its health arm, the World Health Organization (WHO) was established in 1948. That year, the Nobel committee bestowed its prize on Paul Müller. The WHO began to search for a grand project that would, in a sense, legitimize them. Here was the gift of DDT and with it the prospect of ridding the entire world of malaria.

It was no mad scheme; there was logic to it. DDT could be sprayed onto house walls. Many of the major *Anopheles* mosquito malaria vectors would fly inside houses to get their evening meal and once fed to repletion would take their digestive rest on those sprayed walls. The DDT would pass through the chitinous "soles," act on the insect's nervous system, and with trembling shivers, the "DDT jitters," as entomologists called it, the insect would die. Once sprayed, it could be a mosquito death house for as long as six months. With this strategy, DDT antimalaria pilot projects were having spectacular success, particularly in South America under the Venezuelan malariologist Arnoldo Gabaldon. Gabaldon began to lobby for global eradication. In this vision, he was supported by the mathematical projections of George MacDonald, professor of epidemiology at the famed London School of Hygiene and Tropical Medicine. MacDonald's calculations "proved" that if all the walls of all the houses in a malaria endemic region were sprayed with DDT periodically, about every six months, the anopheline vector population would be so reduced that there would be no transmission. In five years, all the malaria infections would be "burnt out," even without chemotherapeutic intervention. Spraying would then be discontinued, except for a few problem foci.

Capital investment in the program would no longer be required, and governments could move on to other health needs. Post-campaign mosquito populations would rise again, but a mosquito without malaria is only a damn nuisance, not a health menace. Malaria would be eradicated from the entire world. After a little palace politicking in which the WHO secretary general, a Canadian psychiatrist who had remarked that "one cultural anthropologist is worth more than one hundred malaria teams," was replaced by a new Brazilian Prince of Malaria, the Global Eradication of Malaria program was launched in 1955.

I won't go into all the problems that led to the demise of the eradication program. Socrates Litsios, (*The Tomorrow of Malaria*, Wellington, New Zealand: Pacific Press, 1996), a WHO insider says that WHO was too inflexible in its prescription to deal with a diverse set of epidemiological problems. A problem not recognized in WHO's Swiss chancery was that many houses in the tropical villages don't have walls. There was also a bit of dissimulation in not acknowledging that the sub-Saharan area was not do-able and China was politically off limits. Such admissions would have made it difficult to get contributions such as the $250 million from the United States if it was advertised as the "Global Eradication of Malaria Except for Africa and China Program." Moreover, human biting (anthropophilic), house resting (endophilic) mosquitoes became progressively more resistant to DDT, and there was growing resistance on the part of the parasite to the cheap, nontoxic antimalarial chloroquine. In 1972, WHO threw in the towel and admitted that the brainless mosquito had proved more cunning, more adaptable than all the mega brainpower of the malariologists.

While the malaria teams were applying comparatively

small amounts of DDT on house interior wall surfaces, farmers were pouring it onto their crops. And who can blame them? For all the 10,000 years of agriculture, farmers have fed almost as much (and sometimes more) of their produce to the insects as they have to the humans. It has been a long struggle to deny the vegetarian arthropods their free lunch. The farmer has always welcomed anything available that would control insect depredation of his crops. Genetic modification, unquestioned under the guise of selective breeding, has long been one strategy, but in the twentieth century, insecticides have been the farmer's big weapon. After World War I, daring young men flew their "Jenny" biplanes, spraying insecticides that included nicotine, derris root powder, lime, sulfur, and the Bordeaux mixture of copper sulfate and lime. Delta Airlines began life in 1930 as a crop-dusting company.

The marginal farmers of Africa and Asia however didn't have a Delta, or any other, crop duster. Their insects chomped away and in bad years, as when the locusts came in overwhelming hordes, they and their people starved. To the small farmers of the third world and the big acre farmers of the entire world, DDT was like a gift from the gods. It was incredibly lethal to insects, long lasting, and so very cheap. No airplane crop duster was required; DDT could be sprayed by a worker with a backpack pump container. By the early 1950s, DDT had produced an immense increase in crop yields, especially cotton. Where people had once gone hungry there was now a surplus of food that could be sold, the new cash source paying for schooling, health services, and improved living conditions.

In the early 1950s, farmers the world over noticed that they had to use more and more DDT to get any protective effect. Predatory insects and their selfish genes were striking back by developing resistance to the insecticide. Spraying crops is obviously unlike house interior spraying; the runoff from rain and irrigation were depositing ever increasing amounts of DDT, polluting the environment. Nevertheless, through the 1950s, there was no serious challenge to DDT. It was just too useful in agriculture and public health. Besides, study upon study had proven DDT harmless to humans. Cancer had been the first concern, but there was no evidence of any increase in any form of cancer in exposed populations or even in spraymen who mixed and applied DDT. There was no evidence of any mutagenic effect, no allergenic effect. Yes, DDT did accumulate in the fat (DDT is fat soluble; in water it is one of the most insoluble substances known) and in mother's milk, but again there were no detectable toxic effects.

DDT turned from saint to sinner in 1962 with the publication of Rachel Carson's environmental manifesto, *Silent Spring*. The book appealed to the conscience of all nations that the environment was in grave danger and nature must be rescued. DDT became the symbol, the "fall guy" for all the chemical pollutants infecting the environment. Clearly, DDT was adversely affecting certain wildlife. The osprey and peregrine falcon were near extinction. DDT had made their eggs so fragile that they cracked before hatching. But that these assaults on the environment had also flowed from agricultural excesses never emerged in the discussion. The spraying of houses or the dusting of people for public health purposes never killed an osprey. The environmentalist wardens did not make that distinction (nor have they yet), and there was no voice of authority to come to DDT's defense.

DDT did indeed have its problems with nature. In Malaya for example, DDT eliminated a wasp predator of the caterpillars that ate the *atap* (palm frond) roofing of houses in the *kampongs* (villages). The wasps died, but the caterpillars did not, and the roofs fell in. My old friend, Dr. Ungku (Prince) Omar, director of the Institute of Medical Research in Kuala Lumpur, came to visit my department in Singapore in approximately 1962 and was telling us how the Malayan malaria eradication scheme was collapsing because the villagers refused to allow their houses to be sprayed. There was also an increase in bed bug infestation because DDT selectively killed their ant predators.

For the environmentalist "Greens," *Silent Spring* came at an opportune moment; ten years earlier, they would have been dismissed as nut cases, flower children, and malcontents. But in 1962, there were few defenders of DDT; the agricultural chemical industry actually welcomed the opportunity to abandon DDT. It was too cheap, and newer, much more expensive and profitable pesticides were ready for the farmer's market. The chemical industry mounted an aggressive political offensive. A chilling article by John H. Cushman Jr. in the *New York Times* (March 26, 2001) sums up the industry's "crash program" to counter the adverse publicity as environmental devils. The industry campaigned to soften state and environmental laws. They paid lobbyists with their PACs to "upgrade the Congress." They formed dummy pressure groups with misleading names like Citizens for Effective Environmental Action Now. And, of course, they fed money to the money-hungry congressmen. Giving up DDT made the industry look self-righteous, that they were working for the best interests of humans and nature. There was no dissent when, in 1972, the newly created Environmental Protection Agency (EPA) banned the use of DDT. America, the main supplier to the world, shut down production of DDT completely.

But who needed DDT? WHO's Global Eradication of Malaria Program was on the rocky road to failure. The WHO laid the blame on the incompetent natives who couldn't or wouldn't follow the strict spray protocol, resulting in the vector anophelines' resistance to DDT. Other apologists laid the blame entirely on the farmers whose constant, intemperate, and high level use of DDT led to the selection of resistant mosquitoes.* Some contemporary scientists even question whether DDT resistance was the pivotal factor that led to the collapse of the eradication campaign. Instead, they blame WHO's arrogant management and technical incompetence for the failure to roll back malaria into extinction.

Some strains and species of mosquitoes had developed an undoubted physiological resistance to DDT. Under DDT "pressure," there is a selection of mosquitoes whose DNA encodes for the amplified production of the "breakdown" enzymes, esterases and aldehyde oxidase. But even DDT resistance may have a bright side. There was a remarkable observation made by L. McCarroll of Cardiff University and his coinvestigators in Sri Lanka that insecticide resistance may be epidemiologically beneficial. The worm parasites (*Wuchereria bancrofti, Brugia malayi*) cause lymphatic filariasis that may, in time, lead to the gross disfigurations of elephantiasis. Transmitted by mosquitoes, it is a major health problem in Sri Lanka where the vector is the sewer-breeding, filthy-water-loving mosquito, *Culex quinquefasciatus*—familiarly known as "quinks." Sri Lanka has been under a cloud of

* The eradication program as a global exercise was a failure but was successful in parts. Malaria was eradicated, and the eradication maintained, in the more-advanced endemic regions such as Crete, Sardinia, Taiwan, most of the Caribbean, and Europe. Millions of lives were saved in countries like India and Sri Lanka during the years of effective DDT spraying.

DDT and other insecticides for over fifty years in attempts to control malaria, filariasis, and dengue, to name a few, and some strains of quinks have become highly resistant. The McCarroll group discovered that the resistant quinks were less capable of transmitting the filarial worms than the insecticide-sensitive strains. Their explanation is that the high level of esterases in resistant mosquitoes kill, or inhibit the development of, the ingested microfilariae to the infective stage. They also suggest that DDT-resistant anophelines may, for the same reason, be inhibitory for the malaria parasite's development within them. It would certainly be worth exploring—if you can still find an able and willing entomologically trained malariologist. Our hoary joke has it—the only thing that the Global Eradication of Malaria Program eradicated was the malariologist.

The other type of resistance to DDT is known as behavioral resistance. Once the anathema of malariologists engaged in control programs, it is now considered by some as a good thing, an additional strength of DDT's malaria-fighting power. As a doctoral student at the London School of Hygiene and Tropical Medicine in 1948, I learned my malaria epidemiology and its statistical foundations from the master himself, Professor George MacDonald. We, his disciples, were instilled with the DDT dogma of the day. Mosquito flies into house. Good. Mosquito lands on DDT-sprayed wall. Good. Mosquito dies. Good; the only good mosquito is a dead mosquito. Many, many mosquitoes die, and there is no more malaria. Finis. There was another dogma for bad behavior. Mosquito flies into house. Mosquito bites. Mosquito is irritated by DDT, buggers off, and flies out of house. Malaria continues.

My entrenched notion of behavioral resistance was to be questioned at the May 2000 workshop meeting, The Contextual Determinants of Malaria, held in Lausanne, Switzerland. The meeting was organized by Carnegie-Mellon University faculty and generously funded by ExxonMobil; its purpose was to evaluate the effects of global warming on malaria (emissions, you know). There were several old-timer participants, veterans of the malaria wars in many parts of the world, and at one session, talk turned to DDT—the value it had been, the value it might still be. A younger participant lauded DDT not for its killing qualities but for its excito-repellent action. This surprised me, because it seemed like behavioral resistance by another name. The new rationale, based on sound experimental data, was that mosquito populations have evolved that find DDT exquisitely offensive. These mosquitoes avoid DDT with the same distaste that you might idiosyncratically be repelled by some substance that you find repugnant, durian fruit, a make of perfume or aftershave lotion. The good of behavioral resistance has this logic: Anopheline mosquitoes bite at night. People sleep at night. People sleep inside their house. Female mosquitoes (you remember, only females are hematophagous—it takes blood to make babies) must, perforce, fly into the house to get their late evening blood supper. However, if DDT is offensive-repellent to the mosquito, it will either not enter the sprayed house or exit before biting.

A leading investigator of excito-repellency is Dr. Don Roberts of the Uniformed Services University of the Health Sciences in Bethesda, Maryland. He and his colleagues studied *Anopheles vestipennis* and *Anopheles albimanus*, important malaria vectors in Central and South America. The normal behavior of both species is to feed inside the house. One of Roberts's studies in Belize compared the behavior of *Anopheles vestipennis* in unsprayed huts, huts sprayed with the synthetic pyrethroid deltamethrin, and huts sprayed with DDT. In their normal behavior (unsprayed huts), they began entering the huts after sunset, bit all night, and began exiting five hours before sunrise. In the deltamethrin treated huts, there was a 66 percent reduction in numbers, and they exited five hours earlier than normal. Nevertheless, there were still enough mosquitoes indoors for a two-hour dinner to transmit malaria as well as to acquire the infection from any human carriers. The DDT-sprayed huts had an astounding 97 percent reduction in indoor numbers of *Anopheles vestipennis*; one whiff of DDT, and away they would go. Roberts et al. concluded, "The repellency effect documented in the DDT-sprayed house essentially excluded human-vector contact within that house. Reduced levels of mosquitoes entering and biting will strongly reduce the potential for malaria transmission."

Despite the militancy of the environmentalists, DDT has not become an extinct insecticide. While its use in agriculture has been greatly reduced, it remains an important weapon for controlling malaria and other vector-transmitted diseases. Since those diseases are not neatly compartmented, the cessation (or resumption) of DDT targeted for one vector can have a cascading effect on another vector. For example, in 1970 India decided to disband their national malaria eradication program. DDT household spraying against anophelines ended. During the good spray years, not only had malaria mortality diminished to near zero, but so had infection and mortality rates from another vector-borne disease, visceral leishmaniasis (known as kala azar, the Mogul for black sickness) transmitted by the *Phlebotomus* sandflies. The sandflies, tiny biting midges, were, if anything, more susceptible to DDT than the malaria mosquitoes—and remained so even when the mosquitoes became increasingly resistant. When spray operations stopped, the mosquito population bloomed and epidemic malaria returned. The sandfly populations, which had been reduced to near nontransmission levels, also rose, and by 1975 there were renewed epidemics of kala azar in India, Bangladesh, and Nepal.

Despite bans and opposition, DDT remains in production, manufactured in India and China, and according to one source, in Mexico. Pakistan may be another producer according to a kala azar control officer I met in Patna, India, who once told me that the DDT he was using was "rubbish" because it came from Pakistan. He noted that it would, of course, be potent if it came from India. DDT and the Kashmir problem? These sources are in danger. Amir Attaran, the malaria advocate lawyer now at Harvard's Center for International Development reports that Greenpeace is trying to shut down the Indian DDT factory: ". . . they may accomplish through the back door what they couldn't accomplish through the front door." I have made inquiries (no response from WHO) but can find no comprehensive inventory where DDT is being used or, importantly, what vectors are still susceptible to it. We really need that catalog to promote the public health case for DDT. Ethiopia, evidently, still uses it in household spray operations. Tanzania has been experiencing pyrethroid resistance in its impregnated bed net antimalaria program and has returned to focal DDT spraying.

Chris Curtis, the London School of Hygiene and Tropical Medicine's comprehensive expert entomologist, affirms that *Anopheles gambiae*, a champion malaria transmitter, is susceptible in Tanzania where it "goes doggedly indoors and commits suicide." There is a malaria epidemic in Burundi with 2 million registered annual cases among the 6 million Burundians. They spray. In South Africa, DDT brought malaria under control in their most endemic province, KwaZuluNatal, so in 1995 they stopped using it. Malaria reappeared. In 2000, there were ten times as many cases as in 1995. They sprayed.

For the worldwide malaria confederation, the Internet is our jungle drum, transmitting messages that would rarely, if ever, be divulged in the starchy medium of science journals. Through an exchange of e-mails, we came to know of Indonesia's plight in its need for DDT to stem a severe outbreak of malaria in central Java. A malariologist-parasitologist expert with experience in Indonesia expressed his concern regarding the potential malaria epidemic disaster in a "densely populated area in the midst of economic and political turmoil." He recommended DDT spraying to contain the outbreak. He has had the agreement and support of the senior responsible official in Jakarta's Ministry of Health, Dr. Umar Achmadi, but they have been frustrated by the total ban on DDT stemming from Suharto's dictatorial days. There is no recorded documentation for this sweeping action but the unofficial story is that it was done to appease the World Bank and the United States Agency for International Development (USAID). There is also a story that one or two shipments of shrimp and/or tobacco to the United States was found contaminated with DDT, and DDT was banned for even medical purposes to protect the investment deals of Suharto's cronies. Finally, the nonmedical United Nations agency, the Environmental Program (UNEP), entered Indonesia into their registry where DDT was to be banned for *all* purposes. The desperation of central Java changed nothing, but this time the cronyism was between UNEP and that other United Nations agency, the World Health Organization. The WHO attempted to pressure the Indonesians to exclude residual spraying, although there was no promise that they would supply the new anti-malarial drug, Coartem, at $54 per case. Achmadi was not intimidated. It was reported that one of the WHO staff, an American posted in Thailand, "chastised" him for "confusing" the Indonesians.

UNEP continues to be the front for all the misinformed, misguided environmentalists who reject DDT's utility in saving lives and who would put *any* environmental advantage over *any* human health need. UNEP fronted for the final assault of the environmentalists led by the World Wide Fund for Nature in concert with Greenpeace, Physicians for Social Responsibility, World Wildlife Fund, and the International Pesticides Elimination Network. In December 2000, the UNEP convened an international meeting in Johannesburg, South Africa, to enforce a worldwide ban of twelve substances (including DDT) they considered to be environmental pollutants. Their concessions to reality would allow exemptions for public health applications if the supplicating nation jumped through all the UNEP hoops. Also, realizing that alternative "pollutants" and strategies might not be around the corner, UNEP decreed that the total ban would not be imposed on the signatory nations until 2007.

The sister agency, the WHO, a party to these proceedings, did what they do best, forcefully equivocated. Several years ago, WHO embarked upon another antimalaria program less ambitious than the 1950s global vision. This was their Roll-Back Malaria initiative which has a goal of reducing malaria-caused mortality in half by 2010. To many who examined their plans, or lack of them, it seemed to be yet another WHO money-gathering exercise to support the Geneva palace and its courtiers. Now WHO had to appease the environmentalists and for this, they said UNEP was correct, insecticide spraying should be ended by the 2007 deadline. In their November 28, 2000, press release, WHO lauds Belize, Costa Rica, El Salvador, Guatemala, Honduras, Mexico, Nicaragua, and Panama for seeking ways of reducing reliance on DDT. No mention is made how this can be done; where DDT is applied, WHO says it must be done according to their strict set of rules. It's déjà vu; that same inflexibility led to the collapse of the (semi-)Global Eradication of Malaria Program of a half-century ago. There were candid and ignored entomologists and malariologists in WHO even during the Eradication era. Some wrote of their opinions then, others waited until retirement to air their dissident views. A WHO sanitary engineer in the Division of Malaria Eradication, H. A. Rafatjah wrote a document in 1972, a time of low reputation for DDT's effectiveness, "Operational and Financial Implications of Replacing DDT in the Malaria Eradication Programme." He concluded that even then there was nothing to replace DDT. "The annual cost of insecticide used in residual spraying of the malaria eradication programme might increase from approximately $20 million to $106 and $400 million if DDT were replaced by malathion or propoxur respectively. Similarly, the total cost of spraying may be raised from about $60 million to $184.5 million and $510 million respectively." Keep in mind, these are 1972 dollar figures. Nothing has changed. There is no practical replacement for DDT, nor is there likely to be in 2007, 2017, or 2027. When a poor tropical country runs into epidemic malaria trouble, they have no recourse but to resort to DDT household spraying to save their citizens.

Undoubtedly, to their surprise and chagrin, the UNEP coterie were to meet a protest from a powerful coalition of scientists and physicians around the world, including three Nobel laureates and the Harvard economist, Jeffrey Sachs. The union boss who organized the "malaria march" was Dr. Mary Galinski, founder of the Malaria Organization. Mary is a noted malaria researcher at Emory University. She is a dynamic force majeure. While still at New York University with her husband, the malaria researcher Dr. John Barnwell (now at the CDC in Atlanta), she began her mission to bring malaria—its status and its needs—to public and professional attention. She singlehandedly created the Malaria Organization, which has since flourished and has grown in strength and services. Its website (www.malaria.org) is an excellent portal of entry to information.

A letter from the Malaria Organization and a coalition that called itself Save Children from Malaria Campaign, signed by 371 scientists and physicians from 57 countries, told UNEP that they had better think things through because "the relevant question is not whether DDT can pose health risks (it does not), but whether those risks outweigh the tremendous public health benefits of DDT for malaria control." What UNEP/WHO are asking the poor, malarious countries to do is wean themselves from DDT (an apt aphorism because the DDT suspension looks kind of milky; on the other hand an outraged New Guinea tribe once thought it looked like seminal fluid) and adopt alternative measures, such as impregnated bed nets and chemotherapy for all who need it. That's good, says the malaria coalition, but it would be immoral to do so until the rich nations give the money to buy those mosquito nets ($5 per net and up) and the new antimalaria drugs to treat the multidrug-resistant strains ($5 per case and up. Way up). Even then, when there is an epidemic emergency, there is nothing else but to take DDT out of the closet and spray.

The pervasive mystique of DDT's evil lives on. Al Gore wrote in the commemorative introduction to *Silent Spring*, "Because Carson's work led to the ban on DDT, it may be that the human species—or at least countless human lives, will be saved because of the words she wrote." Ronald Bailey, *Reason* magazine's science correspondent replied, "Sadly, it's more likely that, because a blinkered orthodoxy cannot accept the heretical notion that DDT has some beneficial purposes, countless human lives will be lost." And exercising the author's privilege to have the last word . . . "Al, I've been a lifelong Democrat. I voted for you. I would vote for you again—early and often if necessary—but you're dead wrong on this one. Think again and get back to me.—Bob Desowitz"

Chapter

4

The Malaria $Millions

Many malaria investigators, especially the well-funded, will disagree with my opinion that one of the more interesting research findings is that the Queen of Vectors, *Anopheles gambiae,* is irresistibly attracted to the vapors of Limburger cheese. This bit of trivia disputes the cheery representations and predictions of the mainstream researchers and their DNA vaccines, genomic roadmaps, Bill Gates's malaria millions, and the World Health Organization's "Roll Back Malaria" slogan. My counter, the cheese effluvium, is offered as a way of saying that, despite all the accomplishments of high science and its big money, there are now an estimated 300 million cases of malaria resulting in 3 million deaths a year. Nothing much has changed, it's even worsened, since our Jewish Grandmother took leave from worrying about her tapeworms in 1982 and told us "How the Wise Men Brought Malaria to Africa."

In 1898 Ronald Ross, the British discoverer of the mosquito transmission of malaria (1897), wrote Alphonse Laveran, the French discoverer of the malaria parasite (1880):

> My dear Dr. Laveran,
>
> I feel however that the malaria problem has been solved in the main outlines. We have only to find the hosts of the parasites in various countries in order to enter upon an era of scientific prevention.

Ross had chutzpah, if one may use that ethnic term about so prideful, starchy a Briton. He may have successfully followed the sporozoite (the infecting stage) to the mosquito, but he was, essentially, ignorant of medical entomology. Ross, and too many of his malariologist successors, did not appreciate the diversity of vector anopheline species. Each species has its characteristic, genetically driven behaviors of ecological habitat and breeding water selection, as well as feeding preferences. There have been successful malaria eradication campaigns where the main anopheline vector bred in large, drainable bodies of water and the "host" nation was sufficiently affluent to undertake a massive engineering project. Two such great antimalaria projects of the 1930s were the Italian draining of their Pontine marshes and the United States' Tennessee Valley Authority project. But even those giant projects of environmental engineering, the Italians called it "bonification," did not completely eradicate malaria. Eradication of malaria happened only when the economic benefits accruing from the projects allowed the population to build better, screened housing, have access to healthcare and such intangibles as improved nutrition and education. And not infrequently, the final step to total eradication required the coup de grâce of a spritz of DDT.

Current DDT debates have been the subject of the previous chapter. Here I would reemphasize that until DDT there was, essentially, no malaria vector control for the poor peoples of the tropical world. For the first time, DDT gave malariologists a weapon to kill a large diversity of anopheline species without the need for environmental engineering interventions. The World Health Organization was established in 1948 and they needed a grand crusade that would, in a sense, legitimize their birth. DDT was then beginning to have spectacular successes in antimalaria pilot projects and some of the grand panjandrums of malaria, Arnoldo Gabaldon of Venezuela being the grandest of all, began advocating global eradication of malaria by widespread DDT house spraying. In 1955, WHO, with a commitment of $250 million from the United States, opened their first and most major ever campaign, the Global Eradication of Malaria. For a time, the program was wonderfully effective (in some regions). Then the mosquito vectors became behaviorally and physiologically resistant to DDT. By 1968, the WHO had admitted that they had lost the campaign.

The foremost malaria repository is sub-Saharan Africa. Even during the Global Eradication Campaign's finest years, nothing could be done to relieve, let alone eradicate, the continent's malaria burden. With the horrifying statistic of an African child's death every thirty seconds from uncontrollable malaria, Africans are further burdened by uncontrollable AIDS, uncontrollable poverty, uncontrollable corruption, and uncontrollable tribal and religious hatreds played out as massacres and wars. We shall return to Africa, but the example of reemergent malaria might better be served by a study of India to convey the sense of proportion and urgency of the current tropical malaria problem.

When criticism was leveled at the WHO for an unrealistic, unworkable malaria eradication stratagem, their defense was "India," an enormous nation with an enormous population at risk to malaria. The landscape epidemiology of Indian malaria ranged through starkly different ecosystems—rice fields, agricultural villages with their man-made ponds ("tanks"), savanna, jungles, hill country, mountain foothills. Each ecosystem had its anopheline species adapted to that habitat. India was and is the only place in the endemic world where there is true urban malaria, transmitted by *Anopheles stephensi*, a species adapted to breeding in the large rooftop water jars. India was the mother country of malaria research. After independence, its scientists have continued the high-quality investigations begun by the British Indian Medical Service.

There had been an efficient, disciplined Indian National Malaria Eradication Program with a professional central authority responsible for its operations. India came so very close to ridding its subcontinent of a disease that since ancient times had crippled its economy and debilitated its peoples. It all fell apart as the mosquitoes became increasingly resistant to DDT. The near half-century of the fall and rise of malaria in India would be, in round numbers, like this:

1948 Independence	75	million cases	800	thousand deaths
1956 National Malaria Eradication Campaign	800	thousand cases	0	deaths
1994 Campaign had failed, Rajasthan epidemic	15	million cases	20	thousand deaths

The resurgence of malaria in India was due in part to the old familiar epidemiological story that well-intended national projects can bring unintended malaria epidemics. Malaria is, to a large extent, a man-made disease; human activities create the ideal breeding habitats so dearly loved and needed by many of the vector mosquito species. Great activities bring great malaria, notably consequent to the internationally financed, massive water impoundment projects for hydroelectric power and agriculture. Examples have been mainly from Africa and undertaken from the 1950s through the 1970s. After that time, these engineering spectaculars had pretty much run their course, although the epidemiological havoc produced has persisted. It was a rich vein for a science writer to mine—but fortunately the hiatus has been filled by the issue of global warming and disease. More on that presently. After 1980, the environmental-ecological lesson had been learned, at least for Africa, and funding agencies such as the World Bank would no longer underwrite these schemes unless their epidemiology experts gave them a "clean bill of health." But the bank made a fatal exception when India decided to make Rajasthan's desert bloom.

Once upon a time, but not too long ago, there were wise men who came to the Great Indian Desert where it was searingly dry, 109°F (43°C) in the summer, underpopulated, and abjectly poor in wealth and goods. Illness often befell them, but even so, in this dry land they were not overly troubled with the fever sickness brought by the mosquito. "Let there be water," said the wise men, and they contrived with the money lenders in the bank for the whole world to give them the millions of rupee lakhs to dig a great ditch to bring the waters from the towering Himalayas at the border to this parched land called Rajasthan. The desert would bloom with food crops and the people would prosper. The mountain to desert ditch they named the Indira Ghandi Canal in honored memory of their motherly leader murdered by her Sikh bodyguards—the very same sect of people who now complained that the Hindus were again abusing them by diverting water from their Punjab homeland. There was now water and crops, but there were also many Anopheles mosquitoes who had come to breed in the lush habitat that the wise men had created for them. From 1994 onward, Rajasthan has been experiencing a severe malaria epidemic. Moreover, with each passing year, an increasingly greater proportion of the cases are of the potentially fatal falciparum malaria. Here is yet another tale of ecological-epidemiological misadventure. I find it disconcerting that the words used twenty years ago for what happened in northern Kenya can be so readily transposed to present day Rajasthan.

Rajasthan's is not the only outbreak; India, once the pride of the malaria eradicationists is beset by a near-nationwide resurgence of malaria. It is difficult to get reliable epidemiologic statistics; on the ground, observers might say, for example, that in their district there have been 4,000 malaria deaths during the year, but the government will deny this and publish a mortality figure of 50 for the same period. Happily, it is a new network world and governments can obfuscate but they can't hide from the Internet. On this worldwide web of informal camaraderie, malaria workers in the laboratories, hospitals, and in the field freely exchange information, opinions, advice, and occasional invective (www.malaria@wehi.EDU.AU). It's all unfiltered, no "decaffeinated" data judiciously edited for peer reviewed publication. During this past year (2001), anguished e-mail messages have been sent to the Internet malaria circle. Two such messages:

> To all concerned:
>
> City of Calcutta, India is now facing a severe crisis from cerebral malaria. Already more than 100 people have died. Is there any international organization which keeps records of malaria-related informations on India and Calcutta? Because local authorities are not willing to share information with the citizen and more busy to hide the facts.

and

> I am serving as the Senior Deputy Director of a 715-bed referral hospital in a tribal area of Orissa, India. We get approx(imately) 800–1000 cases of falciparum malaria patients admitted every year of whom 50 percent are severe. Recently there is a rise in the number of cases of acute renal failure in severe malaria. We are not been able to know the reason thereof. Is it due to a more virulent strain? Can it be due to other drugs? I need your kind comments on it.

A June 4, 2001, e-mail from the Harvard Promed epidemiology surveillance adds, "The death toll in a malaria epidemic sweeping the northeastern state of Assam has risen to 50, and more than 4,500 others are suffering from the disease, officials said Tuesday."

Here is India, a nation that makes atom bombs and can't find a way to control malaria. If India has failed, what can be said of Africa? The great reservoir of falciparum malaria is sub-Saharan Africa. African health professionals are confronted with the greatest challenge—a superefficient mosquito vector, poverty, inadequate health services, wars, and political discord. If a way or ways can be found to control malaria in Africa, then doing so in the rest of the endemic world will be the proverbial piece of cake.

Unlike AIDS, or even sleeping sickness, for which there are no ready answers, malariologists have remedies, stratagems, and theories they assert will solve Africa's malaria problem. There are low-tech advocates who recommend putting cow dung into the small collections of water where anophelines breed. Their technocrat counterparts ask for research funds to genetically engineer a mosquito incapable of transmitting the parasite. The low-tech druggist would have the natives grow their own sweet wormwood (*Artemesia annua*) from whose leaves a decoction of antimalarial tea can be made. The high-tech druggist will try to synthesize a Roundup-like herbicidal drug that will cripple a newly discovered piece of the malaria parasite's enzymatic machinery. And in science's heavens, above the Luddites and the technocrats of drugs and mosquito abatement are their eminences, the vaccinologists. All will be well, they reassure us, the malaria vaccine will soon be here; those nasty drugs and polluting insecticides will soon be history. This has been their message for seventy-five years. There is still no vaccine, but we still believe in them and invest in them because an antimalaria vaccine would be so wonderful. It (the vaccine) is the only game in town, as one high priest and his circle proclaim. Here, I would respectfully disagree. When I, or a Yoruba villager, shakes uncontrollably with a malaria rigor; when parents agonize over their child dying of cerebral malaria, we want that drug. We want it now to save life, not some time in the promised future. We want that antimalarial to act swiftly. It must be made available and it must be made affordable to villager and wealthy traveler alike. On the front line in the war against malaria, *that's* the only game in town!

Drugging Malaria into Submission

Immediate relief could be obtained by supplying some of the newer antimalarial drugs such as mefloquine (Lariam) and artesunate for distribution at primary health centers to the acutely ill. But these are very expensive; a Lariam pill costs almost the same as a Viagra pill. This may not be a difficult choice; I assure you, there isn't much in the way of libido when you are sweating, shivering, and shaking with malaria. In sub-Saharan Africa, where drug resistance has not as yet reached the multidrug dimension of the Southeast Asian *P. falciparum* strains, the relatively inexpensive pyrimethamine-sulfadoxine combination (Fansidar) is still (mostly) effective. But before too long, the African strains of *P. falciparum* will no longer respond to Fansidar, and no one knows what antimalarial drug will be available in tropical Africa when that day arrives. When suitable antimalarial drugs were available, millions of lives were saved among the native populations. Prophylactic antimalarials prevented sickness and death of military personnel and visiting or resident business people. For twenty to twenty-five years, chloroquine had been the most suitable of antimalarials. Even so, except for some pilot projects, there were no extensive mass chemotherapy strategies to eradicate malaria in the manner that DDT in the Global Eradication Campaign was intended to do. When the Global Eradication program collapsed, a turn to primary health services with their ad hoc distribution of chloroquine was advocated.

Primary health care could be as simple as village volunteers working with a minimum of basic training. In one such project in a holoendemic area of Kenya, the volunteers gave chloroquine to those who came complaining of self-diagnosed malaria. Malaria transmission continued, neither parasitologic rates nor serologic titers were affected by the program, but after two years there was a significant decrease in mortality and morbidity.

More than forty years ago, mass prophylaxis was attempted by distribution of common salt medicated with chloroquine or pyrimethamine. Pyrimethamine-salt rapidly led to resistance in areas as widely separated as Cambodia and West Irian. Chloroquine-salt did better despite a tendency for the drug to leach out when stored under conditions of high humidity. Also, there was limited salt intake by young children, the population segment with the highest parasitemias. Nevertheless, we tend to forget today how effective the strategy was in some pilot trials. Here is a report from a "salt trial" in Tanzania from 1962 to 1966.

> Chloroquine diphosphate in the form of medicated salt was sold, through normal commercial channels, to the 2200-3000 inhabitants of the isolated settlement of Mto ma Mbu in northern Tanzania for nearly five years. The sporozoite (infected mosquitoes) rate fell from 4.1 to zero, although in this formerly holoendemic malarious area parasitemia very occasionally continued to be found. The irregular pattern of infections suggested, however, that most if not all were being contracted elsewhere except during a brief period in the middle of the trial when some non-medicated salt was used in error. These encouraging clinical results were achieved despite the known lack of salt in the diet of small children, and despite the loss amounting to one quarter of the chloro-

> quine content during storage after the first seven mixes. Because of this loss, the cause of which was found to be leaching of chloroquine in salt not sufficiently dehydrated before mixing, the actual amount of chloroquine base being ingested each week in the diet of some Mto ma Mbu adults may have been as little as 128 mg and not the 160 mg which the mixing concentration of 0.3% was expected to yield.

We should reconsider this once promising strategy and perhaps try it again with newer antimalarials less likely to give rapid rise to resistance.

Antimalarial drugs will be needed, as they have been for over 400 years, to cure illness and save lives. But what drug(s) fulfill this need today and what drugs might be developed in the future? Except for a few limited areas, *P. falciparum* is chloroquine resistant throughout the world's endemic regions. In vitro experiments have shown that this resistance can be reversed by agents such as Verapamil, a calcium channel blocker, and in vivo in the owl monkey by Desipramine and other tricyclic antidepressant drugs. This line of research does not appear to be pursued, at present, in any major fashion. However, if a cheap, nontoxic synergistic resistance-reversing agent could be found, it would go a long way in solving the current control-by-treatment dilemma.

Despite the synthetic wizardry of modern pharmacology, the final therapeutic arbiter for severe malaria is a botanical derivative in use for some 400 years—quinine. It may well be that the antimalarial to replace quinine in this new millennium will also be a botanical derivative of even more ancient lineage than quinine—artemisinin, the Chinese qinghaosu from the sweet wormwood (*Artemesia annua*). Over two thousand years ago, in 168 B.C., qinghaosu was mentioned in the Han dynasty "Recipes for 52 Kinds of Diseases" as a sovereign remedy for hemorrhoids. Five hundred years later, in 340 A.D., Ge Hong, the author of "Handbook of Prescriptions for Emergency Treatments," advised its use for fever and chills. From that time, it has become a staple of Chinese traditional medicine for that purpose. In 1967, scientists of the People's Republic of China began a systematic search for new drugs from among traditional herbals and discovered artemisinin's striking activity against malaria parasites.*

The active principle of artemisinin is a sequestrene lactone peroxide. It has been synthesized for oral use (artesunate) and an oil-soluble form for intramuscular injection (artemether). Both have remarkable antiparasitic activity, reducing the parasitemia by about 10,000 fold within 24 hours (against a factor of about 10 for mefloquine). However, artemisinin has a short half-life, requiring a course of five to seven days—a well-tolerated regimen with little side effects. This course of artemether brings rapid parasitological and clinical resolution in children with severe and cerebral falciparum malaria. For uncomplicated hyperparasitemic falciparum malaria, a course of oral artesunate is effective, especially when accompanied by a single dose of mefloquine or three doses of pyrimethamine-sulphadoxine.

The artemisinin antimalarials are not approved by the United States FDA, but they are manufactured elsewhere, by Rhone-Poulenc and by a company called Arenco. There is a

* There is another wormwood plant, *Artemesia absinthum,* from which the now illegal liquor, absinthe, is made. In the 1840s, the French government gave it to their soldiers stationed in Algeria, in the mistaken belief that it would protect them from malaria. In the light of artemisinins, I wonder if that belief was completely mistaken.

product called Arsumax produced in China by the Guilin Pharmaceutical Works for the French company, Sanofi Winthrop. The artemisinins are excellent, almost ideal antimalarials, but their cost, about $6 for the five-day course of treatment in sub-Saharan Africa, where the total malaria budget of a country usually does not exceed $.50 to $1 per person, may make it unavailable to the people who need it most.

China's newest formulation of an Artemisia derivative may be the most effective malaria curative known. It is now sold under the trade names Riamet and Coartem. If you have severe falciparum malaria and are European or Chinese and have $57 in ready cash to buy the curative dose, then Riamet is the drug for you. It is marketed by that monster-sized Swiss pharmaceutical company, Novartis. In 1996, Novartis absorbed two semi-monster-sized Swiss companies, Sandoz and Ciba. At $61 billion it was the biggest corporate merger in history.

Chinese pharmacologists had spent twenty years trying to improve the antimalarial action of the *Artemesia* extracts. By 1990, the Institute of Microbiology and Epidemiology of the Academy of Military Science in Beijing had come up with a formulation, a combination of artemether and the synthetic, benflumetol. This preparation which they designated as Co-artemether had the highest activity yet observed in their experimental screen. The choice for the Chinese government was whether to give it to the world's malarious millions or to exploit it for the malaria $millions. Well, each according to his needs, as the Prophet Karl said, and in 1990, the Chinese government's Science and Technology Committee approached Novartis with an offer to codevelop Co-artemether.

Novartis was uncertain; they had no experience with

malaria or any other tropical disease. However, several years earlier, they had set aside a pool of company money for altruistic projects in the developing world—which might eventually also earn them, with any luck, some speculative Swiss francs. They called it their Risk Fund and with it had underwritten such diverse undertakings as the recycling of cattle tick dip, agricultural advisory services in Bolivia, and export promotion of Senegalese textiles. Surely, an antimalaria joint venture with the Chinese was as worthy as recycling cattle tick dip. A million Swiss francs ($726,000) was allocated from the Risk Fund to support the further testing of Co-artemether's effectiveness and safety. It proved to be better than the best expectation. Over 95 percent of falciparum cases, even those with multidrug-resistant strains, were cleared of their parasites and fever within thirty-six to fifty-two hours with little or no adverse drug toxicity. With these results Novartis, decided to commercialize Co-artemether. In 1994, Novartis entered into a joint licensing agreement with their Beijing pharmaceutical partners. A Swiss patent was granted and Co-artemether was ready for sale under its new, Novartis name, Riamet.

There was no altruistic intention to give, literally "give," Riamet to the malarious millions in Africa and other endemic developing regions. Novartis and its Chinese partners expected big profits from sales to tourists and other rich, mostly white, Europeans who dare venture where the anopheline flies.* Their commercial expectations were supported by changes in personal antimalaria strategies.

* Riamet is not licensed in the United States nor has Novartis any intention of applying to the FDA for approval. Artemisinin derivatives also have not been screened by the FDA. For Americans, the new "stand-by" antimalarial is Glaxo Wellcome's Malarone (atovaquone plus proguanil).

In the happy days of chloroquine-sensitive strains of *Plasmodium falciparum*, you could go to your general practitioner and tell him that you were going to a Kenya game park or that you had a job with an oil exploration company in Nigeria. He would prescribe chloroquine pills to be taken once or twice a week. That would be sufficient protection—it was a prophylactic (and at a higher, more prolonged regimen, it would be curative if you came down with falciparum malaria). Then, beginning in about 1972, chloroquine resistant strains appeared and spread to become a global condition. Chloroquine's prophylactic place was taken by Roche's Lariam (mefloquine), an effective, albeit much more expensive drug. Millions of people, including myself, have taken Lariam as a prophylactic (it's not all that good as a curative) without any untoward side effects. However, in a relatively small percentage of people, Lariam has been associated with neurological disorder—which is a roundabout way of saying that Lariam made them quite wacko. Most of the Lariam loonies have been Europeans, but more and more people are avoiding it as an antimalarial prophylactic. The Internet has an anti-Lariam group, complete with photographs, giving testimony against Lariam. Those who refuse to take Lariam must adopt other, time-honored, defensive strategies. They are advised to take evasive action—wear long-sleeved clothing in the evening, apply repellants to skin and clothing, sleep under a mosquito net or in a screened room—and carry that "stand-by" $57 bottle of Riamet pills should the malaria parasite break through these defenses.

Riamet's life-saving values were not lost on the health authorities trying to deal with Africa's malarious millions. In "normal" times, impoverished Africans would have reluctantly accepted the way of the world in which the rich got richer and healthier while the poor got death and disease. The African AIDS holocaust has ended that passivity. At first, they accused the multinational pharmaceutical companies with their $2,000-a-year AIDS drugs as being heartless bloodsuckers. When this didn't do much good, the Africans threatened to ignore patents and arrange "pirate editions" of the drugs they needed to keep them alive. The African AIDS Revolution was about to blow away that whole shaky house of patent protection cards. The alarmed drug companies quickly conceded and arranged for either giveaways or very much lowered prices.

Malaria is second only to AIDS in impact on the health of Africans; some experts argue that, over time, it is more important. The desperately needed new antimalaria drug, Riamet, was there but, like the AIDS therapeutics, it was too costly to be available. In the new revolutionary spirit, Novartis got the "AIDS treatment." The WHO, acting as an agent for the Africans, met with the Novartis executives and asked for a deep discount. Novartis knew that already several small commercial laboratories were exploring the feasibility of producing Riamet with—or without—the legitimacy of a license. In May 2001, Dr. Daniel Vasella, Novartis's chairman, announced that for Africa the cost of a curative course of Riamet would be $2, their bottom-line cost of manufacture. Novartis was willing to forgo profit but not lose money. Curious mathematics—if $2 covers manufacturing cost, then there would be a $55 profit when sold at the drug store.

David Alnwick, WHO's malaria manager, expressed the organization's gratitude but pointed out that $2 was a lot of money for the average African, especially when compared to chloroquine's $.20 treatment price. There was a wringing of hands; the Nigerian president Olusegan Obasanjo asked that all the continent's foreign debts be forgiven so they could pay for AIDS-malaria needs. Money is now pouring into Africa for those projects. The American fiscal year 2002 budget for global health is $1.3 billion. Kofi Annan has asked for $8 billion; President George W. Bush offered $200 million, and Congress responded by introducing a bill to give $700 million more. The World Bank has anted up $300 million to $500 million as an interest-free "antimalarial" loan. The Bill and Melinda Gates Foundation is donating $115 million for malaria; most will go toward research but some is to be allocated for purchase of antimalarials. Contributions by other countries, such as Japan, and nongovernmental organizations have not been tallied but they must be sizeable. There should be enough money to buy enough Riamet pills to cure every African child, woman, and man of their severe malaria; malaria mortality could be reduced to zero. But will antimalaria pills get to those who need them? Will the bulk of that drug money end up in the parasites or in the accounts of parasitic politicians, national, international and nongovernmental apparachiks, and research scientists?

A small, but vocal group have espoused a novel counterbalancing stratagem to the issues of big drug technology, big international administration, and big money. They propose a do-it-yourself sweet wormwood plan as a way of circumventing oppressive pharmaceutical neocolonialism. The group's informal leader is a doctor in Belize, Peter Singfield, who often couches his e-mail messages in socialist dialectic. These do-it-yourselfers propose a way to liberate the natives from both malaria and the drug companies: give them seeds of *Artemesia annua* which they can raise in their garden plots. A German group, the Anamed Coordination Group, have developed a high-yielding hybrid, *Artemisia annua anamed,* that grows well in central Africa where they have tested it on several hundred people with severe falciparum malaria. They claim the curative results to have been excellent. They also distribute seed kits, and the plants now flourish in Cambodia, Uganda, Tanzania, Sudan, and South Africa (see www.anamed.org). For the herbalists among my readers, here is their recipe and regimen: One liter of boiling water is poured onto five grams of dried leaves of *Artemisia annua anamed.* It is allowed to brew for ten to fifteen minutes, and then poured through a sieve. The tea is then drunk in four portions in the course of a day. The period of treatment is between five and seven days. I don't know what happens if you smoke it.

In Cloud Cuckoo Patentland *Artemesia annua anamed* remains "uninvented." Will it survive independently in the present practice of patenting the botanicals of nature? The most recent plant skirmish has centered on a tree that grows in India, *Azadirachta indica.* In 1994, the U.S. Department of Agriculture and the multinational agribusiness, W. R. Grace received a patent on fungicides extracted from the tree's seeds. The Indians indignantly claimed that they had been using the same extracts as fungicides for decades and the tree's products for cosmetic and medicines for centuries. There are ninety patents worldwide based on *Azadirachta*'s products. To this the Indian's and Bangladeshi's reply is "biopiracy." Theirs has been an emphatic reply complete with demonstrations—500,000 strong in Bangladesh. Europe's Green Party has also expressed strong indignation and confronted with this scrutiny, the European Patent Office has revoked, disinvented, the fungicide patent. Only eighty-nine more to go. One of the tree's extracts has now shown, in the experimental screen, to have significant antimalarial action. If confirmed, it will be interesting to follow the fate of the tree's juices as an antimalarial and patentable product. Oh yes, *Azadirachta indica*'s familiar name is the neem tree which is the Hindi for "free tree."

I don't want to dismount the botanical biopiracy-patent law hobby horse before telling you of the Tumeric Trials. Tumeric is one of south Asia's traditional spices that is also believed to have medicinal properties. In the 1950s, Indian researchers demonstrated that tumeric, topically applied, promotes wound healing. Yet, in 1993, the U.S. Patent Office awarded the University of Mississippi Medical Center patent #5,401,504, "Use of tumeric in wound healing." The Indian Council of Scientific and Industrial Research cried "foul" but Ole Miss has, so far, retained the patent. Tumeric hasn't been tested for any antimalaria activity—but who knows? My personal worry is that my favorite dish is the ambrosial Indonesian *rendang*—small chunks of beef sauteed with tumeric, ginger, garlic, lemon grass stalks, and chili peppers and cooked to dryness in coconut milk. Will I need a license from the University of Mississippi Medical Center before making my next batch of *rendang*?

For the moment, the derivatives of artemether, with the potentiating help of some synthetic compounds, should be capable of curing all of Africa's malaria. The problem will be in the future when *Plasmodium falciparum* will become resistant to the artemethers, as it has, over time, to every other drug thrown at it. Prudence and experience dictates that a battery of new antimalarial drugs must be held in reserve. The impediments are: (1) the major pharmaceutical firms are reluctant to carry out the expensive research and development for these nonprofitable drugs, and (2) novel approaches must be found to invent these new classes of antimalarials.

The development problem was addressed in 1999 when the WHO established, as a semiautonomous, not-for-profit Swiss foundation, the Medicines for Malaria Venture (MMV). It has been funded by the WHO, World Bank, Bill and Melinda Gates and Rockefeller Foundations, ExxonMobil Corporation, and the Swiss, Netherlands, and British governments with a yearly budget of $30 million. The MMV has invited antimalaria drug research proposals from pharmaceutical and biotechnology companies, big and small, as well as from individuals in university or research institutions. The interest is certainly out there; in 2000, over one hundred letters of proposal were received from twenty-seven countries. That first year, $4 million was allocated, each awardee having some connection with a major pharmaceutical company.

But the MMV hasn't planned for success. What will happen if one or more successful antimalarials emerge from the MMV? Where will the big money be found to bring the new antimalarial to market at an affordable price? What actually is that cost? The drug industry, when attacked for the high price of their products, trot out the figure of $500 million as their cost to bring a new drug from laboratory to pharmacy. That figure has never been convincing. True, the drug industry is a notorious big-spender. If it takes $900 bottles of wine to interact with clinical investigators (your physicians) then that astronomical sum might be accurate. As far as research costs go, the public contribution has never been accurately factored. A 1997 study by the National Science Foundation reported that 73 percent of the published papers cited in patent applications described publicly funded research.

The second problem—where will the new classes of antimalaria drugs come from?—is being met by new and exciting research on the parasite's biology. These studies have yielded findings that may lead to rational drug development. As an example, here is a study indicating that for those with malaria, Roundup may be in their future medicine cabinet.

Malaria parasites belong to a group of protozoans, the Apicomplexa, which have intricate subcellular structures (organelles) at their anterior end. In *Plasmodium,* two of the apical organelles contain protein ligands that when expressed are crucial in the invasion of the red blood cell. Two other organelles support the concept that eukaryotes, organisms whose chromosomes are within a discreet nucleus, not only evolved but were also *assembled.* It has been known for some years that the mitochondrion of eukaryotes, the laminated organelle that is the "battery" performing chemical energy-producing transactions, is actually a bacterium engulfed by some protozoan ancestor. That bacterium defied digestion and survived over the billions of years, dividing when its host cell divided. It became a permanent paying guest contributing, symbiotically, to its host's life process.

And so eukaryotes (creatures with nucleated cells) contain DNA in both the chromosomes within the nucleus and extranuclearly in the mitochondrion. In addition, *Plasmodium* and *Toxoplasma* (another Apicomplexan pathogen of humans) have DNA of a third kind within a membrane of an organelle called a plastid. Plastid DNA had been considered to be nothing more than than a break-off fragment of mitochondrion DNA, but in 1997, a study by Sabine Kohler and her colleagues revealed, astonishingly, it to be a relic of a chloroplast DNA from a green alga incorporated, like the archaic mitochondrion-bacterium, by some ancient Apicomplexan ancestor.

The comparable plastid DNA in plants and fungi encodes seven enzymes that mediate a biochemical process known as the shikimate pathway in which erythrose 4-phosphate is converted to chorismate. Chorismate is essential in the production of *p*-aminobenzoate (PABA). PABA is then converted into some of the amino acid building blocks of life. Mammalian and other animal cells do not have a shikimate pathway and get their PABA exogenously. That's why a powerful herbicide, glyphosphate, which inhibits an enzyme in the shikimate pathway will kill the weed and spare the gardener. In June 1998, Fiona Roberts and Craig W. Roberts and their coworkers reported that *Plasmodian falciparum* and other apicomplexan parasites have a shikimate pathway presumably encoded by the circular thirty-five kilobase plastid DNA. Moreover, the growth of the malaria and other apicomplexans was reported to be inhibited by the herbicide glyphosphate.

The chloroplast-*Plasmodium* connection intrigues researchers engaged in discovering new therapeutic antimalarials. They now see the possibility of attacking the parasite at its botanical side—a malarial herbicide, so to speak. Earlier, in 1997, Maria E. Fischera and David S. Roos of the Department of Biology, University of Pennsylvania, reported that the fluroquinolone antibiotic, Ciprofloxaxcin, is a potent inhibitor of *Toxoplasma* replication by specifically attacking the plastid DNA. Fischera and Roos suggested that fluoroquines, acting in similar fashion, may be among the next generation of antimalarials.

The Mosquito Millions

For one hundred years, beginning with the apostolic preaching of Ronald Ross, a basic tenet of malaria control has been the interruption of transmission through the anopheline vectors. When DDT and other insecticides began to lose power, alternative, mostly biological, means of control were sought. Essentially, there are three strategies for the control of malaria transmission: (1) the mosquito numbers are reduced to the point where little or no transmission occurs; (2) the mosquito is somehow prevented from biting its human blood supply; and (3) the infection in the mosquito is prevented, in which case the mosquito is merely a damn nuisance but no longer a malaria vector.

A diversity of biological, relatively low-tech methods have been applied to mosquito control. High-technology mosquito control remains mostly in the minds of the scientists. Some are very simple "home remedy" methods like the Russian Sochi method of putting cow dung in small collections of breeding water, such as footprints or rutted roads, to kill the larvae. This, of course, only works where the anopheline vector breeds in those kinds of puddles and you have a cow to make the dung. Then from India comes a recommendation to grow "tulsi," holy basil (*Ocimum sanctum),* which is claimed to repel mosquitoes. Larva-eating fish, notably the guppy (*Poecilia reticulata*) and the mosquito fish (*Gambusia affinis*) have been introduced into waters where vectors breed. Germ warfare, deliberately spreading viruses, bacteria, and fungi pathogenic to the mosquito, has been attempted. There have been demonstration trials in which the larva-specific toxin and spores of *Bacillus thuringiensis israelensis*/serotype

H-14, a bacterium isolated in 1976 from Israel's Negev desert, has been cast onto breeding waters.

The antimosquito measure currently receiving most attention is the very old-fashioned mosquito net but now fortified with pyrethroid insecticide/repellent. Pyrethrum, the natural product of the chrysanthemum flower, has been used as a "knockdown" insecticide for many years. Its limitation is "staying power"; it degrades rapidly. In the 1970s, synthetic pyrethroids were developed. These, particularly permethrin, were not only potent but had residual activity for as long as six months. Beginning in the late 1980s, large scale pyrethrum-impregnated bed net trials have been carried out in Africa and Asia. It has been a kind of "home remedy," with villagers being taught how and when to dip/redip their nets.

In each trial, after a year of usage there was a dramatic 30 percent or more reduction of overall mortality in children under five years. In fact, the results were better than they should have been; the impregnated bed nets prevented more disease than was accountable from malaria. The reduced contact with mosquitoes and other insects engendered by permethrin may, in some unknown manner, have a preventive effect on nonmalarial causes of mortality such as diarrheas and respiratory infections. More likely, malaria has been underestimated as a direct or contributing cause of child death.

But there are reservations about the impregnated bed net. It is possible, but still unproven, that the paradoxical long-term effect is actually increasing malarial mortality. In populations subject to numerous bites of infectious mosquitoes, the children under five years of age are most likely to be affected by malaria. They have the highest rate and intensity of parasitemia, the most fever, the most profound anemia, and under certain endemic conditions, the most cerebral malaria. Slowly, as they grow older, a functional immunity intervenes by a mechanism(s) imperfectly understood. And as they get older, they begin to live more or less compatibly with their malaria parasites. The exception is the first-time-pregnant women who lose their hard-earned immunity. One school of malariologists has long held the belief, “don’t mess with Mother Nature”; for them, childhood death is an acceptable price for a stable, productive, malaria-tolerant adult community (but of course, the children of the malariologists aren’t being culled). There are concerns that the impregnated bed net will interfere with the development of immunity, that susceptibility to severe malaria will ultimately be shifted to older and older age groups.

It had been assumed that with a high, constant assault by the infectious mosquito there would be the consequent risk of severe, fatal, or near-fatal malaria in the under five-year-olds, but again the enigmatic parasite confounded the malariologists. When a British-Gambian-Kenyan-American group of researchers led by Robert W. Snow of Oxford University compared childhood hospital admissions for severe malaria in African communities under heavy, moderate, and light endemicities, they found that it was the moderate transmission situations that make for malaria’s killing fields. Death from cerebral malaria and severe malaria struck children of about two to three years of age in settings where transmission intensity was about half of that of the “heavy” endemicity settings. In villages of light endemicity, less than one-tenth of the heavy-moderate force of infection (a mathematical formula which derives a relationship between age and exposure to *P. falciparum* parasites), there was less than 2 percent parasitemia in children one to nine years of age and very few hospital admissions for severe malaria. In view of these data, the concern has been raised that in these clinically stable settings of very high transmission, intervention methods such as impregnated bed nets will depress transmission only to the moderate "killing field" level rather than to the very-low transmission-safe level.

Impregnated bed nets are saving lives, but for modern malaria researchers, this is only a stopgap to a technical innovation that will cripple the anopheline's biological transmission machinery. Today, this high technology is confined to the thoughts and the laboratories of the scientists, but these great expectations are worth looking at.

An aspect of malaria's evolution would nip the parasite "in the bud"—to disallow transmission by making the mosquito biologically refractory to infection. Bird malarias are transmitted by culicine mosquitoes. The malaria species infecting mammals, whether they be mice, mouse deer, monkeys, or men are transmitted exclusively by anophelines. The moment in time that the mammal, the anopheline, and the *Plasmodium* came together to form their interdependent, cyclical unit is unknown, but over millennia, that exclusivity has become absolute. Put a *Plasmodium* of mammalian malaria in a culicine mosquito and the parasite dies; put a *Plasmodium* of a bird or lizard in an anopheline mosquito and the parasite dies. Malaria researchers have long been perplexed by the Anopheles-Culex barrier, but to understand it may be one way to control malaria. With no new operating methods to reduce mosquito vector numbers, the alternative of interruption of transmission by preventing infection of the mosquito has been sought. The exploitation of a refractory mosquito's genetically determined factors seemed impossible before the gene transfection methodology, which has opened possibilities for an enormous, diverse array of biomedical and bioagricultural benefits—including malaria control.

Taking a page from the successful control of the screw worm by massive release of sterile male flies, male mosquitoes, reared in "dud stud farms" and rendered sterile by irradiation or chemosterilants, have been released into the environment. These measures have been successful in varying degrees but none, so far, have proved sufficiently practical or effective to be adopted in large-scale antivector campaigns. Even so, the distinguished malaria researchers A. Warburg and L. H. Miller consider the mosquito-inhabiting stages of the malaria to be the most vulnerable points of attack. But they also point out that it is really a complex chain of weak links. Experiments with different pairs of malaria parasites of humans and animals and their refractory mosquitoes have revealed that there is no one single stage; the parasites' progress is variously halted from first-step male gamete formation (exflagellation) to last-step sporozoite.

These are exciting times for the malaria researchers. Those attacked by malaria do not share that enthusiasm. One can only hope that before too long they will have the practical rewards from the laboratory. Perhaps the great immune hope—a malaria vaccine.

Chapter

5

Malaria: Millions for the Vaccine but Not One Cent for Defense

A malaria vaccine has, for the past seventy years, been the rational answer to the malaria problem. With an effective vaccine, neither the costly, constant search for new antimalarials (that too soon become ineffective and/or are too costly for those in greatest need) nor the arduous, too costly antivector operations would be necessary. So why not a vaccine? After all, we have vaccines against all sorts of bacteria and viruses. Smallpox was totally eradicated with a vaccine. And it's not that humans don't have the immune "right stuff." Those who live in stable, endemic areas do have the capacity to mount, eventually, a functional immunity. Certainly twentieth-century science should be able to improve upon feeble nature. It is not from want of trying. The long, long trail to the malaria vaccine began in 1910 and still has not arrived at the often-promised destination. Every trick in the immunologist's book has been applied. Hundreds of millions of dollars have been given to the research. Chickens, rats, mice, and monkeys have been successfully immunized with vaccines made of *their* malaria parasites. Only humans, oddball mammals, cannot (yet) be protected in an acceptable, practical manner by a vaccine made from *our* species of malaria parasites. Why?

First of all, there is the impediment of stage specificity. A bacterium is a bacterium and a virus a virus, but a malaria parasite is a many splendored thing. When I was still in the business of sacrificing graduate students, it was the custom to put them through the life-cycle catechism. It would go something like this: the sporozoites from the mosquito to the liver where they develop into pre-erythrocytic schizonts, maturing to merozoites, which burst out to enter red blood cells, maturing progressively from ring form to trophozoite, to schizont, to merozoite, which burst out to invade new red blood cells, etc., etc. That's the asexual cycle. For the sexual cycle, there are the male and female gametocytes in the blood, becoming male and female gametes in the mosquito, which "mate" to form an ookinete, which becomes an oocyst in which sporozoites develop. Don't worry about this strange-sounding terminology; get a tropical medicine text if you are *really* interested.*

The malaria-inherent life-cycle rule is that each one of the phases given as stage names not only is different in morphology but also distinct in its antigenic makeup. Each has its own immune-inducing signature. Thus, if a vaccine is made of whole sporozoites (the stage in the salivary glands of the mosquito that initiates infection when the mosquito feeds)

* There is a cartoon drawing explanation in my earlier book *The Malaria Capers* (W. W. Norton, 1991). Some of the early history of malaria and the malaria vaccine is covered there.

or an antigenic component, it has to induce an absolutely solid immunity because there is no cross-protection with the disease-causing blood stages; any breakthrough would be as if there was no immunization at all. Conversely, immunization with an asexual blood-stage (schizonts or merozoites) vaccine would affect parasite numbers and disease but would have no impact on transmission through the mosquito. On the other hand, a vaccine made of the mosquito's sexual stages could halt development/transmission in the mosquito but would have no effect on preventing disease.

The second big obstacle to immunization is the malaria parasite's skittish antigenicity encoded by a large family of skittish genes. Many years before the advent of modern molecular immunogenetics, malaria workers knew that people exposed to constant bites by infected mosquitoes developed a functional, clinical immunity usually by the time they reached the early teens. It was also observed that when these semi-immune individuals moved to another endemic area—it could be a village only a few miles away—they would lose their immune protection and come down with a severe malaria attack. Clearly, the "local" immune response is limited by parasite variation. With the elucidation of the molecular-genetic mechanisms underlying antigenic variation, we are coming to appreciate how the plasmodium is truly a master of evasion and disguise.

Three species of human malaria (*Plasmodium vivax, Plasmodium malariae, Plasmodium ovale*) can be debilitating, but only the fourth species, *Plasmodium falciparum*, can kill, and it is the primary target of vaccine development. The old, "classical" attempts to make a malaria vaccine were based on the fact that for the successful bacterial and viral vaccines, whole organisms had been attenuated, inactivated, or killed.

Current vaccine philosophy has adopted the genetic mystique. This proposes that one or more parasite proteins control the functional immune response, or with great good fortune, a solid, sterile immunity.* These proteins can be isolated from the parasite; better still, their encoding gene is isolated and by recombinant technology inserted into a bacterium or yeast which is duped into making the wanted protein. Even better, the antigenic segment of the protein is isolated and its amino acid chain sequenced. The chain can then be manufactured from amino acids taken from the stockroom shelf to make a vaccine that had never known a DNA parent.

A major candidate vaccine antigen (All are "candidates" until elected by a successful trial in humans. There have been many candidates, but we are still waiting for the Elected One) is a *Plasmodium falciparum* protein, or proteins, that is deposited on the outer membrane surface of the infected red blood cell. This surface protein(s) is essential to the survival of the malaria microorganism in its cat and mouse game of frustrating the human's immune system. The protein(s) functions as a ligand, anchoring in lock and key

* Functional immunity in malaria seems to depend on the coexistence of low numbers of parasites. This has been termed "premunitive immunity." In contrast, a solid, sterile immunity is such that the pathogen is absent and the immune status so powerful that infection is prevented. The actual nature of that immunity remains unknown. One school holds that it is only acquired when the person has had antigenic experience of many variants, and that is why it takes so long to develop. Another school holds that it is a function of age; the young have an immature immune system, and that is why it takes so long to develop. A third, less popular school (in which I am enrolled), holds that there is a prenatal immune priming (PIP) in which, at a critical period of fetal development, the mother by cells or antigens sensitizes—primes—her fetus to be born with an already defending immunity. If our school is correct, it would explain why some babies die and some babies survive. A vaccine may then work best, if this is true, by immunizing the pregnant mother and then immunizing the infant.

fashion the infected red blood cell to the molecular receptors on the surface of the cells lining the blood vessel walls. This allows the *Plasmodium falciparum* to hide from the general circulation (it is the only one of the human malarias to have deep vascular sequestration), while it is maturing within the red blood cell. This keeps it from harm's way—the activated, parasite-killing immune cells of the spleen. A vaccine that would turn the immune response against these sequestration antigens would, theoretically, shift the advantage to immunity's "cat." But this parasite is one smart "mouse." *Plasmodium falciparum* has a large family of genes, maybe 6 percent of the entire gene constitution (the genome), that responds rapidly to the production of specific antibodies. Each time an antibody is produced against the parasite's sequestration antigen, another gene is turned on to encode for a new antigen amino acid sequence unrecognized by the antibody. Eight or ten variants (clones) can circulate in the infected individual at the same time. The magnitude of *Plasmodium falciparum*'s repertoire is not known with any certainty, but these genes, the *var* genes, can produce thousands, if not millions, of antigenic variants. Unfortunately, other candidate vaccine antigens may have the same lability.

Then there has been the difficulty in the vaccine antigens not having sufficient "punch" to yield the desired kind or level of immune response. After all, even the real thing, the living malaria parasite, induces a functional immunity only after many years of stimulation. There are, however "vaccine helpers," adjuvants, which can boost an otherwise poorly performing antigen. That adjuvants seem to be an absolute necessity to most, if not all, nonliving malaria vaccines has been known for sixty years. World War II opened without adequate antimalarial prophylaxis, either by drug or anti-vector means, to protect Allied troops in endemic zones. Research to develop a vaccine was then considered to be a worthwhile war effort. Jules Freund and his colleagues made the major effort, using avian and primate model malarias. They discovered the second, enduring principle/barrier of malaria vaccinology: a blood-stage vaccine could be effective but only when used with a powerful immune booster, an adjuvant. That adjuvant was the oil emulsion-killed *Mycobacterium* (a bacterium of the tuberculosis family) mixture known as Freund's complete adjuvant. It fostered protection but it produced side effects so dire that it could never be used in humans. In the almost sixty years since Freund's experiments, many vaccine preparations of all parasite life stages have been developed, but in each instance there remains that pivotal barrier, the need for an adjuvant booster. The variety of adjuvants tested reads like an alchemist's shopping list but none, with the possible exception of the hemolytic, plant-derived detergent, saponin, and its newly isolated fraction, has been able to replace Freund's adjuvant.

It is of historical interest to note here that the German armies were also having malaria problems and, in their own monstrous way, attempted to develop a vaccine. Between 1939 and 1945 the plasmodium took a heavy toll on German troops in Crete, Greece, and North Africa. Continuous protection, chemoprophylaxis, with atebrine was proving difficult. A vaccine was an attractive alternative, and they thought they had just the right man, Dr. Klaus Carl Schilling, to deliver it to them. From 1933 to 1941 Schilling had worked in Italian hospitals in Siena and Voltera where induced malaria was used to treat syphilitic paresis. During that time, he'd made vaccines of parasitized blood attenuated with quinine and tested them in his paretic patients. None of his vaccines were protective, but Schilling falsified the results. He informed the German embassy in Rome of inducing complete immunity in his paretics and was now ready to do the same for the *Wehrmacht.* On December 4, 1941, Himmler ordered Schilling to Germany to begin trials with his vaccines. Schilling now had a seemingly inexhaustible supply of human guinea pigs—the prisoners of Dachau. He began with 200 Polish priests and newborn babies. Seventeen are known to have died. Next in line were the Dachau Jews. As many as 2,000 were used in this way and as many as 500 to 1,000 were murdered by malaria. The meticulous Germans issued a death certificate for each victim, all died of a heart attack. Schilling was captured, stood trial at Nuremberg, and on May 28, 1946, at the age of seventy-two years, he was hung for war crimes. I think that hanging was too good for him; justice would have been better served by giving him malaria and having him experience the death of his victims. There was no Nazi malaria vaccine.

World War II closed with chloroquine and DDT in the antimalaria armamentarium. Interest in the vaccine waned, only to be resuscitated twenty years later when those two antimalarial instruments began losing power. A major revivalist was the United States Agency for International Development (USAID) whose occasionally troubled program has been funded for almost thirty years.* The search

* For those readers of *The Malaria Capers,* published in 1991, and written before the legal conclusions of the cases of those AID-funded researchers accused of stealing money from their malaria vaccine grants, here is what happened. James Erick-

was joined by the British and scientists from other nations. However, from 1968 to about 1980, the studies were, essentially, repetitive of Freund. During this time, the merozoite emerged as a prime vaccine candidate—but it too, without Dr. Freund's adjuvant tonic, was powerless.

By 1980, modern genetic-immunology technology, was beginning its explosive evolution and it had a profound impact on the direction of malaria vaccine research. Whole parasites were out; molecular subunit vaccines were in. The molecules and their controlling genes of the parasite's critical life processes continue to be identified and isolated. Some of these molecule-antigens have been "manufactured" by recombinant bacteria and yeasts, others have been synthetically assembled from "off-the-shelf" chemicals. The surface ligand molecules and apical organelle products of merozoites responsible for attachment and invasion of the erythrocyte have been especially favored as vaccine candidates. So, as noted earlier, have the ligand molecules that bind the *P. falciparum*–infected red blood to the vascular endothelial receptors.

Another candidate immunogen is the small, simple repeat of amino acids of a peptide enveloping the sporozoite. This vaccine would be designed to abort the pre-erythrocytic

---

son, the AID administrator of the program, served his time and is now living a nonmalarial life somewhere. Miodrag Ristic was dismissed from the University of Illinois. He died shortly afterwards. Days before Wasim Siddiqui was to stand in criminal trial brought by the State of Hawaii, he pleaded no contest. After a weekend in jail, he served his sentence by wearing one of those electronic monitors around his ankle. He continued to come to work each day at the University of Hawaii's School of Medicine. The university's professors' union successfully defended his assertion that he not be dismissed because there was no just cause. After his sentence expired (six months, I think), he was allowed to retire with a pension. However, on almost the day of his retirement, he did not contest a civil suit brought against him for some $250,000.

cycle in the liver and thus prevent the subsequent blood infection.* Then there are the controversial sex antigens—the transmission blockers—a group of peptides on the gametocytes. Antibodies raised against them by a vaccine would only work within the mosquito where they would kill the "naked" sexual stages, the gametes or their zygote (ookinete).

Despite their elegance and rationality the subunit vaccines have been bedeviled by the same limitations confronting the whole-organism-vaccine hunters of fifty years ago. Those relatively few subunit vaccines tested in humans or experimental animals still demanded a potent adjuvant to turn them on. The other perpetual problem is that these antigenic peptides are as stage-specific as the life-cycle stage from which they are derived. This has led to a sometimes spirited stage-specific advocacy. The researchers working with the blood-stage peptides claim their vaccine should have priority because it would prevent morbidity and mortality. The researchers working on sporozoite antigens maintain their's would be the ideal vaccine because it would prevent infection. The orthodox malariologists hold that the *real* antimalarial strategy is the interruption of transmission and that this could best be accomplished with a transmission-blocking vaccine.

The military, mindful of their malaria losses in successive wars and operations, have been ardent pursuers of the vaccine. Their protection needs are different from those of

* To recapitulate: The sporozoite in the mosquito's saliva enters the blood stream and within about thirty minutes travels to the liver where it enters a liver "building-block" cell. There it divides repeatedly to produce thousands of merozoites. After one to two clinically silent weeks, the merozoites are released into the bloodstream to begin the red blood cell–infecting stage—with the onset of fever and other pathology of malaria.

third-world communities in endemic foci. Adult, well-fed, disease-free soldiers, not poorly nourished African children beset by other infections such as intestinal parasites, would be the recipients. Military campaigns are of relatively short duration and a six-months protective period afforded by a malaria vaccine would be acceptable. For the army and navy, a second-class vaccine would be good enough for government work—and this is, so far, what they got. A vaccine was devised by a group of military researchers, most of whom were at the Walter Reed Army Institute of Research. The vaccine they devised was produced by the pharmaceutical firm SmithKline Beacham Biological of Rixensart, Belgium. It consists of two *Plasmodium falciparum* circumsporozoite peptides of 207 to 395 amino acids recombinantly expressed in yeast. This recombinant product is fused to a viral hepatitis B surface antigen recombinant product. The parasite-viral antigen was, in turn, combined with a potent adjuvant, one component of which is a saponin fraction, QS-21. Thirty years ago, saponin, from the bark of the *Quilla saponaria* tree, had been shown to be a powerful adjuvant in an experimental malaria vaccine. However, because of its potential hemolytic properties it was never used in any vaccine preparation for humans. A commercial research laboratory, the Cambridge Biotech Corporation of Worcester, Massachusetts, isolated a saponin fraction they named QS-21, which seems to have all the potency of the crude product but is nontoxic when tested in humans. This QS-21 adjuvanted vaccine completely protected six of seven volunteers against the infectious bite of falciparum-carrying mosquitoes. The vaccine has now undergone a phase II trial in the Gambia. Two hundred and fifty adult men received three immunizing injections over five months. Immunization was 65 percent effective for the first two months after the last shot—then it lost virtually all its protective properties.

There is also a group of malaria vaccinologists, led for many years by the Naval Medical Research Center's Captain Stephen L. Hoffman, M.D., who believe that the only method to circumvent such problems as adjuvanticity and antigen immunogenicity is through the very modern DNA genetics methodologies. Their concept, which has worked most successfully with experimental virus vaccines, is to insert by two alternate routes a segment of the malaria parasite's genome encoding for the desired antigen/immunogen into a living carrier. One method is to make a vaccine by inserting the malaria parasite's DNA into a nonpathogenic virus such as the vaccina virus. The altered virus insinuates itself into the host cell genome, which then elaborates the antigen. The other route is to put the desired plasmodial DNA segment into a plasmid and inject the plasmid intramuscularly into the human. The plasmid-parasite DNA is thereby transferred to the muscle cell, myocyte, genome, which, in the genetic sense, now becomes the plasmodium parasite transcribing and elaborating the desired antigen; theoretically, for the life of the new "host."

The asserted advantages of a DNA vaccine are: (1) There is a constant antigenic stimulation which should lead to a higher state of functional, long-lasting immunity. (2) The parasite immunogen so elaborated would contain a much greater array of the antigenic, stimulating segments, the epitopes, than that in a subunit or synthetic vaccine. This would make it a superior, more immunogenic vaccine because (a) there is a "greater sum of its parts," and (b) it would contain the various epitopes necessary to switch on all the different immune systems/components believed necessary in functional immunity to malaria. The DNA vaccine may, in fact, be the best way to switch on the cytotoxic T-cell ($CD8^+$) immunity that has now been shown to kill the parasite-infected cells of the liver. (3) Importantly, a multistage DNA vaccine can be produced in either a single or multiple virus/plasmid carrier. This would cover all the bases, with immune barriers at the sporozoite/pre-erythrocytic, erythrocytic, and mosquito (transmission-blocking) host-parasite interfaces. The DNA vaccines are projected to be easy and cheap to manufacture, and so stable they do not require a cold chain for delivery.

Problems and reservations regarding DNA vaccines have also been voiced. Some of them are: (1) The introduction of parasite DNA into the human genome might, somehow, be cancer-causing (oncogenic). (2) The needed epitopes in their exact chemical construction might not be transcribed by the genetically altered myocyte. (3) Long-term, continual antigenic stimulation might induce immunopathologies such as immune complex diseases of malaria (e.g., nephrotic syndrome), hypersplenism, immunity-associated anemia of inappropriate erythrocytic destruction, and hypersensitivity.

Nevertheless, DNA projects for human trials are progressing and are in progress. The Naval company has a vaccine with DNAs that will express ten antigens, five from the sporozoite/liver stage to turn on the cytotoxic killer cells and so prevent any infection, and five from the red cell–inhabiting stages to reduce the clinical severity of any breakthrough infection by switching on the immune arm that makes the antibody. With his Hoffmanian wit, Steve calls this the MuStDO-10 vaccine, which is an acronym for multistage malaria DNA-based vaccine operation. The DNA vaccine has now been tested in healthy human volunteers and has proved safe, and tolerable—and "suboptimal." It just lacks the excitation of the $CD8^+$ T cells that was expected. The new strategy calls for a DNA vaccine injection followed by a booster shot of a conventional vaccine preparation with adjuvant. So, in a way, we are back to square one. In 1997, the Hoffman group announced in a somewhat obscure journal that a clinical trial with the DNA vaccine had commenced. In November 2000, over a convivial late-night steak and wine dinner in Houston, the company was informed that it was a done deal; four-year-old children could now be protected from malaria. But there has been nothing in print to confirm this except that a trial in semi-immune adults was planned to be carried out in Ghana in 2001 or 2002. To complicate this tale of Hoffman, Steve has seen greener genes on the other side and retired from the U.S. Navy to join Celera. Where MuStDO-10 and other Navy vaccine research will now go, if anywhere, without its long-time leader is unpredictable.

Much the same DNA strategy is being tested in semi-immune adults in the Gambia by the University of Oxford group led by Adrian Hill. Their vaccine, injected intramuscularly, is also the DNA encoding for surface antigens from sporozoites and liver stages. It too is followed by a "conventional" vaccine booster.

Thus, very few candidate preparations have come to phase I or II trial in humans and fewer still to the litmus test of a phase III trial. Very few researchers have had the resources or nerve to test their vaccines in humans. Bold ventures accompanied by big money attract the daring—pirates, politicians, and scientists. The resurrected drive to discover a malaria vaccine with its infusion of grant money and promise of Nobel-deserving fame has attracted, felons aside, some "interesting" people. One of the most "interesting" of the new malaria prospectors has been the Colombian biochemist-cell biologist, Manuel Patorroyo. His has been the only vaccine to be tested in thousands of humans in a range of endemic settings, and it is worthwhile to give a fuller account of his dogged search for the vaccine.

To my regret, I have never personally met Manuel Elkin Patorroyo. I would probably like him but would also probably disagree with him. Some who do know him describe him as charismatic, driven, brilliant, and of high social sensitivity. Others who know him speak of him in less flattering terms. Patorroyo certainly has the best of scientific credentials; in the midst of his medical school studies in Bogota, he was awarded a Rockefeller Foundation scholarship and left to go to Yale as a graduate virology student. He then transferred to the Rockefeller University where he earned his Ph.D. in 1984. The now Doctor Patorroyo returned immediately to Bogota where he established his vaccine-dream/dream-vaccine laboratory. In a July 24, 1999, interview with Paul Brown of the *Guardian*, he is quoted as saying, "Since I was eight I have dreamed of making vaccines, particularly one for malaria. I am not interested in making money—in Colombia being rich makes you a target anyway. I am prepared to make the vaccine and sell it at cost. The drug companies will not like that and I will face opposition, but think what difference it will make to the world."

The world-saving vaccine Patorroyo is referring to is the still somewhat mysterious Model II. Model I was also heralded as world saving but, novel as it was, proved to be a dud. Patorroyo reasoned that if combined as a chimaeric multistage vaccine of asexual and sporozoite synthetic constructs, it would act in a broadly effective manner. "Synthetic" was the revolutionary novelty; the amino acid sequences for what was believed to be the immune-stimulating portion of the proteins from the sporozoite surface coat, the merozoite surface coat, and internal schizonts (dividing stage) was known. Patorroyo made these sequences (peptide) by stringing the amino acids together in their proper order. This man-made malaria vaccine he called SPf66 and showed it to be safe in Aotus monkeys (the South American owl monkey, the lower primate susceptible to experimental infection with *Plasmodium falciparum*) and nine Colombian soldiers. By 1986, he was prepared to inject it into thousands of people.

The wisdom of the day held that large-scale malaria vaccine trials in humans was near impossible. A World Bank economist estimated that each Phase I (safety) and Phase II (larger-scale efficacy) study would cost between $300,000 and $1,200,000. Distinguished and prudent malariologists worried about money, safety, rules of conduct, rules of interpretation, and sites of studies. The "freewheeling, flamboyant, quixotic, and mercurial—with a penchant for hyperbole" (as the January 20, 1995, issue of the journal *Science* described him), Patorroyo ignored the Nervous Nellies and did it his way—on a grand scale. During the next five years, between 1988 and 1993, he injected 32,000 men, women, and children in Colombia, Venezuela, Brazil, and Ecuador, with the immunizing course of SPf66. It was an incredible, triumphant effort; while researchers in the north were struggling to get a few people tested with their vaccines, here was Patorroyo in the south testing *his* vaccine on 32,000 people at risk to malaria. But did it work? Was SPf66 the vaccine solution to global, particularly African, falciparum malaria?

By the Patorroyo accounting method, SPf66 was approximately 35 percent effective. Others were either doubtful or had difficulty in interpreting the measure of "effectiveness" as percent reduction in symptomatic malaria. There were none of the conventional baseline statistics or controls that mathematically strict epidemiologists demand. Patorroyo had magnanimously assigned all SPf66 rights to the World Health Organization, and the World Health Organization returned the favor by declaring the vaccine to be a significant advance in the continuing fight against malaria. I gave Patorroyo's publications describing his methods, data, and conclusions to my students in the infectious disease epidemiology course as an exercise. If there is something you don't quite understand, give it to your students and tell them it's an exercise. Some agreed with Patorroyo's conclusion, some said it was nonsense, and some, like myself, had difficulty in coming to any definite opinion.

Dr. Patricia Graves, a statistics-epidemiology expert who had been through the malaria wars in Papua New Guinea, and H. Gelband of the Department of Preventive Medicine and Biometrics at the University of Colorado, amalgamated all of Patorroyo's data for his South American trials and put them through the scrutiny of some high-class meta-analysis (mathematical synthesis) programs. They came to the conclusion that there was indeed an overall protection of about 30 percent, but there was variation from trial to trial; sometimes SPf66 afforded this level of overall effectiveness, other times it did zilch. Why? Population genetic differences? Vaccine batch differences? Nevertheless, Pat recommended that the results were good enough to continue human studies with SPf66.

Two other experts, Sarah Gilbert and Adrian Hill of the University of Oxford, had begged to differ. After analyzing the same data, they wrote, "Recently extensive testing of the candidate vaccine SPf66 led to the disappointing conclusion that SPf66 does not protect against clinical malaria and that further efficacy trials are not warranted." SPf66 wasn't getting any confirmatory good news, even from those who wished it well.* The malaria research group at the Centers for Disease Control in Atlanta cooked up a batch of SPf66 according to Patorroyo's recipe and vaccinated Aotus monkeys according to Patorroyo's regimen. The "immunized" monkeys were then challenged with "hot" *Plasmodium falciparum* and promptly came down with infections that would have been fatal if untreated. The monkeys hadn't even made antibody after the SPf66 injections. Patorroyo sent SPf66 of his own making to Atlanta. More monkeys. The same negative results.

To the critics, SPf66 in South America was a sideshow. The continent had foci of malaria but not enough malaria to meet the challenge of the vaccine (and statistical proof). The defining test would have to be in Africa—in African children. Tanzania, where malaria is fiercely hyperendemic, was chosen as the main show's stage. It was to be a mostly Hispanic operation with the principal researchers from the University of Barcelona and the vaccine itself formulated by Spain's Llorente SA. NANP, the synthetic sporozoite coat antigen was added to the vaccine for good measure to determine if it could prevent new infections by nipping the sporozoite at the mosquito's mouth.

* During the time of SPf's fall from immune grace, it was suggested that SPf was the vaccine of the future but was bad-mouthed by jealous northern scientists who didn't want to see a poor third world country boy (ex-Yale and Rockefeller University) succeed where they failed. Nonsense, said the northern researchers, but in fact they worried that all the resources for human trials were being invested in Patorroyo, leaving disillusionment and an empty trial purse when it came to test their own favored vaccine candidates.

Between 1992 and 1994, 274 one- to five-year-old children were given three shots of SPf66 over a six-month period. At the same time, 312 similarly aged control group children were given "dummy" placebo injections. For a year after the third injection, the malaria attacks in both groups were followed for number and severity. To go to the heart of the data: 109 of the placebo control children got malaria as compared to 79 of the vaccinated children. This was claimed to be a crude efficacy of 25 percent which with a little statistical massaging improved it to 31 percent, a marginally significant protective effect. Moreover, the NANP sporozoite component afforded no additional protection against new infections. Nor did the SPf66 lessen the clinical severity of the malaria attacks in the vaccinated children. Graves and Gelfand cast a new eye on the Tanzanian results and, after stringent statistical treatment, came to the succinct conclusion, "There is no evidence for protection of SPf66 vaccines against *P. falciparum* in Africa."

The end came to SPf66 in the trial that was taking place across the continent in the Gambia. It was conducted by the British Medical Research Laboratories in Banjul. The Laboratories had, over decades, carried out some of the most innovative and informative malaria research. In the Gambian study six- to eleven-month-old children were immunized. These children had not yet experienced malaria, unlike the Tanzanian children who were immunologically "primed" by earlier infections when they were vaccinated. The Gambian infants represented Africa's most vulnerable, the age group that an effective vaccine would have to protect. The immunizing regimen followed that in Tanzania. None of the vaccinated children were protected and, even more discouraging, there was, inexplicably, more malaria among the SPf66-vaccinated. In contrast, about the same time as the SPf66 trial, the investigators in the Gambia were showing a 40 percent protection of infants by the simple use of impregnated bed nets.

If you think that the Gambia was the end of Patorroyo's vaccine dreams, think again. The front page article in the July 29, 1999, *Guardian* reported an interview with Patorroyo in which he announced that SPf66, Mark II, with 100 percent efficacy was on the way. The Colombian government continues to finance the enterprise. The king and queen of Spain have pledged their support—as has Fidel Castro. When SPf66 Mark II comes into play, the Spanish government will give $60 million to vaccinate 60 million African children.

Patorroyo says he has tested his "100 percent vaccine" in an astounding 11,500 owl monkeys. To assuage any conservationist rage, the monkeys, captured in the wild by the local Indians, are returned to the wild, after experimentation, by the same Indians. To get through this obviously grinding schedule, that's a lot of monkeys in what must be a lot of experiments, Patorroyo has a staff of 180 divided to work around-the-clock in two twelve-hour shifts. Behind Patorroyo's Immunology Institute of Bogota, located in the Foundation Hospital San Juan de Dios, there is a piece of land about the size of a football field where he intends to build his malaria vaccine factory.

There have been no published reports about the 100 percent SPf66 since the *Guardian* interview in the summer of 1999, so it is hard to judge how things are progressing. However, the last Patorroyoan event adds another bit of the unusual to the somewhat comic opera scenario of the SPf66 Follies. The February 1, 2001, issue of *Nature* published a full-page article with the title—in big black block letters—

"Bank Raids Malaria Centre in Dispute over Landlord's Debt." The hospital housing Patorroyo's Immunology Institute owed the Spanish-owned BBVA Banco Ganadero $2.5 million. The hospital didn't have the cash to repay the debt. The bank confiscated some hospital property, but it wasn't sufficient repayment to make up the shortfall, so they impounded the Institute's equipment, although the institute was only a renter and had no affiliation with the hospital. The bank took DNA sequencers, supercomputers, and magnetic resonance imaging equipment. The Colombian citizenry were outraged. Students threatened to march on the bank. Letters to the newspapers advised that investors and all who banked with BBVA withdraw their funds. This was sufficient pressure to make the bank reconsider. The equipment was released and, presumably, 100 percent SPf66 is again on the way. Patorroyo and Company went back to their twenty-four-hour-a-day work and we await the 100 percent Mark II SPf66.

But suppose you are a poor gringo malaria vaccine researcher without royal patronage or that from the maximum leader of Cuba. How are you to pursue the vaccine dream? Fortunately the American royal family, Bill and Melinda Gates, the billionaire rulers of Microsoftland, have bestowed their great wealth on improving the health of the tropic's poor.

It seems like only yesterday (ten, fifteen years ago?) when Bill and Melinda were married on Lanai and closed the beaches during their honeymoon—which aroused the righteous indignation of the Hawaiian citizenry. As I write this, we have a house guest from Honolulu, our old friend and neighbor April Ambard. April was recounting how furious she and her doctor-sailor husband were when they were excluded from Lanai's Manele Bay boat harbor. For the Gateses, it must be a good marriage because they established a Bill and Melinda Gates Foundation with an endowment of an unbelievable $21.1 billion—that's billions. By law 5 percent of a foundation's assets must be disposed of each year. The Gates Foundation gives $554.5 million a year to global health and research. This magnificent philanthropy is not prompted by a desire to improve the image of the man who owns a company that has been characterized as a ruthless monopoly; it is a gift to the world by two caring people. I apologize for my Lanai resentment, Bill and Melinda. Take any beach you want. Take Waikiki for your personal pleasure.*

The Gates Foundation has been convinced that a malaria vaccine is achievable and is bankrolling a good piece of the malaria vaccine research. It has given the World Health Organization $100 million to test the illusory vaccines. Last year, it contributed $50 million to support malaria vaccine research. To distribute these funds, a Malaria Vaccine Initiative was set up in Rockville, Maryland, with the Peruvian-born, American-trained pediatrician Dr. Regina Rabinovich as director. The initiative is supposed to establish research policy, select trial preparation sites and protocol, but it is difficult to discern how much independence it really has. To some, undoubtedly overly cynical observers the initiative is a "dummy corporation," and the tune is really called by the scientists of the National Institutes of Health's Institute of

* It is a remarkable coincidence that the two greatest philanthropic benefactors to global health, especially infectious disease research and control, have been named Gates. It may have been Rockefeller's money, but it was the Baptist visionary Frederick T. Gates who was the impetus to his philanthropy. This philanthropy, beginning in 1897, undertook the global hookworm campaign (with the American South as the starting project), established the Rockefeller Foundation and the Rockefeller Institute, which discovered, made, and distributed the yellow fever vaccine.

Allergy and Infectious Disease. A midwife at the initiative's birth was the National Institutes of Health's Office of Technology Transfer to make sure that any vaccine progeny would be properly patented.

The problem with all this largesse and the best of intentions is that there is no malaria vaccine; and it is just as reasonable to believe that there never will be one as it is to believe that it is a done deal only awaiting the inevitable breakthrough. The malario-economists are already calculating the costs. One says that the annual cost of the vaccine will be $750 million—providing the manufacturing industry will sell their $10-a-dose vaccine for $.50. No less a person than the Harvard economist, Jeffrey Sachs, director of the Center for International Development, makes his pronouncements with the malaria vaccinologist's assurance that the vaccine is about to happen. Sachs allows $10 for a course of immunization, which would be given to the 25 million children born in Africa each year. This comes to $250 million yearly, which is only 1.5 percent of the $16 billion in aid given to Africa each year.

Seventy years of unrewarding malaria vaccine research doesn't call for abandoning the search, but it certainly calls for a hard look at what is being devoted to it. The vaccine research is the stuff that the National Academy of Science membership is made of, but this research hasn't saved one child, pregnant woman, traveler, or soldier from severe malaria. There are new ways to combat malaria—new drugs, new impregnated mosquito nets, new understandings of mosquito behavior, new tolerance to spraying residual insecticides—and the vaccine is not, or may never be, one of them. Well thought out, strictly supervised, corruption-free projects combining the effective antimalaria strategies could be undertaken with some of the Gateses' or other philanthropists' moneys. Some of the brightest malaria researchers could be weaned from the vaccine and made to think about dealing with malaria in the untidy world we live and die in. Perhaps we need a malaria authority that can forcefully make policy decisions. It should definitely not be, in my opinion, the World Health Organization—a too-politicized body, best at furnishing slogans.

Chapter

6

The Curious Case of the Wake-Up-from-the-Dead Drug and the Bearded Lady

First loves you never forget, whether the red-haired girl in your fourth-grade class or the first pathogen in your life of microbiology. My earliest pathogen love object was the tsetse fly-transmitted trypanosome of sub-Saharan Africa. It was kind of a love-hate relationship; those fishlike protozoa were, experimentally, a pain in the ass. They were vexing and illogical creatures. Try to make a protective vaccine and the little buggers changed their antigenic type, taking this evasive action every three days, ad infinitum. They were difficult to maintain in test tube culture; difficult to transmit experimentally through the tsetse fly; difficult to treat—many of the drugs were, and still are, very toxic formulations of arsenic.

Arsenic has been the classic poison of murderers and playwrights but it hasn't always had a bad reputation. No less a doctor than the Swiss Philippus Areolus Paracelsus recommended it as treatment for a wide assortment of illnesses, especially syphilis.* But that was in the sixteenth century and, although it remained as a therapy for syphilis until antibiotics took over, it has gradually fallen into disuse. However, despite its toxic shortcomings, it has remained a therapy for the Gambian form of African sleeping sickness.

For many infectious diseases, there was a treatment before a cause was known. Malaria is a familiar example; quinine, the "Jesuit powders," had been given in Europe as a specific antimalarial since the sixteenth century (Jesuit priests in Peru brought it to the very malarious Italy), but the malaria parasite itself wasn't discovered until 1880. Similarly, simple inorganic arsenic compounds were used to treat African trypanosomiases of domestic animals and humans before the trypanosomes themselves were revealed under the microscope by Sir David Bruce in animal *ngana* (a Zulu word meaning feeble or weak) and in 1903 in humans by Bruce and Count Aldo Castellani.† In 1858, the great missionary explorer David Livingstone gave sodium arsenite to his draft oxen as he intruded into fly bush. Sodium arsenite was mentioned in the first, 1898, edition of that perennial bible of tropical medicine, *Manson's Tropical Diseases*, as giving good results in "negro lethargy, or sleeping sickness of the Congo." However, these inorganic arsenic compounds were

* Paracelsus had to contend with the prescriptions laid down 1,400 years earlier by that authoritarian Greek doctor, Galen. Galen had some peculiar ideas, although he did show that the arteries carried blood and not hot air. Galen did not recommend antimony or arsenic, and if he didn't recommend it, then it wasn't to be used. There is evidence that until the eighteenth, and perhaps early nineteenth century, the medical graduates of the University of Paris had to swear an oath never to use antimony and arsenic.

† Castellani was also an English knight, but that knight was different from all other knights because he became a Mussolini fascist, and the knighthood was rescinded. He then reverted to a mere Italian count again.

especially toxic, even lethal, at therapeutic dosages. Only when the German physician-scientist Paul Ehrlich formulated organic arsenicals, could trypanosomiasis be treated with a modicum of safety.

In 1908, the flaky Russian Élie Metchnikoff, the Father of Cellular Immunity, shared the stage and Nobel Prize with the flamboyant yet austere cigar-smoking German Ehrlich,* the Father of Humoral Immunity. However, Ehrlich had yet another claim to science paternity as the Father of Rational Chemotherapy.

The preceding fifty years to that Nobel ceremony had been a golden era of scientific discovery. During this brief period of discovery, the life and physical sciences had coalesced to illuminate the causes and cures of what are now called infectious diseases. In 1870, Louis Pasteur began the enlightenment with his discovery that a protozoan parasite, a microsporidian, was killing the silk worms of France, driving an important industry to near ruin. This established the principles that (a) microorganisms could be pathogenic, and (b) that specific microbial pathogens cause specific diseases. From the microsporidian-diseased caterpillars the principle rapidly expanded to include the bacillus of anthrax, the staphylococcus of boils, the plasmodia of malaria, and the mycobacterium of tuberculosis. The catalog of pathogens continues to expand.

Having put a specific microbial face to a disease, it became apparent why old therapies had such limited efficacy. It explained why, for instance, quinine worked against malarial

* Freud and Ehrlich looked somewhat alike and both were photographed smoking a cigar. Freud was reputed to have said, introspectively, that, "Sometimes a cigar is only a cigar," but Ehrlich was the real chain-smoking stogie afficionado, twenty-five or more a day, a cigar box the constant companion under his arm.

fevers and not all fevers, or why the plant product ipecacuenha worked against *Entamoeba histolytica* dysentery but not all dysenteries. The principle of specific therapy made the pharmacologically minded men of science envisage going where none had gone before. The boldest of those pioneers, the drug lord of his day, was Paul Ehrlich. He envisioned synthesizing the *therapia sterilans magna*, the chemotherapeutic that would cure by a single injection: "We must learn to shoot microbes with magic bullets." In his pursuit of the one-dose drug, he exploited the ancients' use of arsenic and antimony, and the new findings of the rapidly evolving field of organic chemistry.

In the mid-nineteenth and early twentieth centuries, coal powered the Industrial Age. The new discipline of organic chemistry was then beginning to show how dead (e.g., coal) and living life forms were chemically constructed and how those molecules might be extracted or synthesized, sometimes for commercial use. The chemical deconstruction of coal gave rise to the dye industry which, in turn, gave rise to the pharmaceutical industry. In 1834, the German F. F. Runge published a paper on the distillation of coal tar to yield the blue dye, aniline. Twenty-two years later, in 1858, the Englishman W. Perkin took aniline as the starting point for his attempt to synthesize the antimalarial, quinine.

All those imperial colonizers—Britain, Germany, France, Belgium, Holland, Spain, and Portugal—had to consolidate and administer their tropical possessions. That rule would be threatened and diminished if their military, rajs, governors, residents, prefects, and district officers were debilitated with recurring, too often fatal, malarial fevers and dysenterys. Further, a sickly subject population could not provide the energetic workforce to produce raw materials, make roads, buy blankets and muzzle loaders—the activities that the colonial rulers expected of their natives. Malaria was a major impediment for both rulers and ruled. The sovereign antimalarial remedy, indeed, the only antimalarial remedy during the Victorian age, was quinine. Quinine was expensive; the supply from South America uncertain (the Dutch were just beginning their cinchona plantations in Java), and the botanical product difficult to standardize. Even today, quinine is a tough molecule to make "off the shelf" and in 1858, Perkin was bound to fail.* But he started the chemotherapeutic train rolling with the notion that specific drugs could be made to cure specific diseases.

What did come from Perkin's Victorian chemistry set was another coal tar dye of gorgeous purple/mauve. By the end of the nineteenth century, these dyes were used to stain microscopic preparations of tissues, bacteria, protozoa, and other parasites. It was observed that the dyes were selective, staining some organisms but not others, or some parts of an organism, such as the nucleus, but not the cytoplasm. Ehrlich, who thought in terms of "grand unities," had been advancing the concept that antitoxins (later to be shown as antibodies) attacked and neutralized their targets by a specific lock and key mechanism. The antitoxin/antibodies were complex molecules of a characteristic three-dimensional molecular structure which fit into a reciprocal structure of the toxin molecule. Similarly, he thought, the dyes he

* The synthesis was three-quarters done by the great American chemist and Harvard professor Robert Burns Woodward and his colleague W. Doering in 1944. Complete synthesis of quinine was not accomplished until 2001, 150 years after Perkin's first go at it. Gilbert Stork and his coworkers finally figured out the technique to make quinine "from scratch." (See Steven M. Weinreb, "Synthetic Lessons from Quinine," *Nature*, 24 May, 2001.) Also, there is a very readable account of Perkin's life and work in *Mauve* by Simon Garfield (W. W. Norton, 2001).

had used to stain histological tissue and pathogen preparations during his early research career were also complex chemical structures joining to receptors of the pathogens. The dyes were, in effect, colored bullets—but were they *magic* bullets?

Ehrlich had shown that the dye, methylene blue, would stain and delineate the malaria parasites within red blood cells. It only remained to find some malaria patients—not all that difficult in the malaria-endemic Germany of 1890—and inject them with the dye that would make them, literally, blue bloods in order to cure them. In 1891, he and a collaborating physician colleague, Dr. P. Guttmann, injected two malaria patients with methylene blue; it was all fast-track, experimental medicine then, no FDAs, no human experimentation committees. The methylene blue was no magic bullet, but it did work; the parasites slowly disappeared from the blood, and the symptoms of fever and rigor slowly abated. With this, Ehrlich lost interest in malaria.* He had succeeded in synthetically creating an organic compound that could be used to treat an infectious disease that held vast regions of the tropics and, at that time, Europe and the United States, in a vise-like deadly grip. He had done it and now, as my old friend the late biochemist Jock Williamson used to say after what he considered to be a stellar accom-

* Methylene blue never achieved status as an antimalarial but curiously, the Nobelist pharmacologist-physiologist Sir Henry Dale in his introduction to a volume of Ehrlich's collected papers notes that in the 1950s the Greek malariologist Dr. H. Foy told him that he still used it to treat quinine-resistant cases of malaria. Also, other workers kept at the methylene blue molecule and finally turned it into a yellow compound, atabrine, which was the prime antimalarial during World War II. Atabrine turned the skin a bright yellow and was dropped as an antimalarial drug. There is a fascinating new report that atabrine shows promise in treating human victims of Mad Cow disease.

plishment, "Let the hacks take over." However, the patriot Ehrlich was still tuned to Germany's colonial needs and he turned his attention to the trypanosomes of African sleeping sickness.

Until World War I stripped away her colonies, Germany ruled vast areas in sub-Saharan Africa. Tanganyika in the east and Cameroon in the west were German. These regions were also colonized by a still more formidable master, the tsetse fly—bringer of the trypanosome. In the west, they were confronted with the Gambian form of sleeping sickness caused by *Trypanosoma brucei gambiense.* In the east, there was the look-alike *Trypanosoma brucei rhodesiense,* the pair being derived from a third look-alike, *Trypanosoma brucei brucei. Brucei brucei* is restricted to wild and domestic animals, mostly ungulates. It doesn't infect humans because a factor in human serum kills it. Evolution, it is believed, had conspired to produce the mutant *gambiense* and *rhodesiense* types that resist the lytic effect of human sera. *Rhodesiense* has retained its animal affiliations and, in a sense, is still a harmless infection of wild antelope, with accidental spillovers to humans. *Gambiense* is, essentially, a parasite of humans, although reservoirs have been found in domestic animals such as the pig.*

If *Trypanosoma brucei rhodesiense* is likened to the plague bacillus, *Yersinia pestis,* in that it rapidly overwhelms and kills its victims, then *Trypanosoma brucei gambiense* can be compared to the human immunodeficiency virus of AIDS, which causes a slow death. Gambian trypanosomiasis has, typically, low numbers of parasites in the blood stream. Initially, it

* *Trypanosoma brucei gambiense* is something of a geographical misnomer. The range of the trypanosome extends well into central and east Africa. In the Congo, Sudan, and Uganda, it is endemic and epidemic.

causes fever and aches that may be confused with a malaria attack. The real problem lies with the Gambian trypanosome's subsequent progress in the body, first from the blood to the lymphatics, and then to the central nervous system's brain and cerebrospinal fluid. It is in that late stage that the manifestations occur which give sleeping sickness its name. There are intense headaches, bizarre behavior, and a lethargy leading to the final, fatal coma. The late stage also causes the greatest difficulty for those treating the patient.

Cells lining the vessels of the brain's blood supply are our unique guardians, denying passage from blood to brain of substances that normally get to other organs. This is known as the blood-brain barrier. For example, an intravenous injection of a drug could rapidly get to the lung and cure an infection, or another intravenous injection could stop the heart for execution, but they might not cross the blood-brain barrier. Similarly, there are chemotherapies that attack trypanosomes in the bloodstream but are ineffective when the parasites have invaded the central nervous system. Blood or brain, in 1900, there were no really safe, effective trypanocidal agents for humans.

Drug discovery at the turn of the century was largely empirical. Hundreds of thousands of compounds were screened for activity against the target microbe or other pathologic condition. It is not all that different today. From a starting molecule that shows some activity, an oxygen-hydrogen OH group can be added here, a nitrogen-hydrogen NH group there, another benzene ring stuck on, and so on. This empirical screening obviously can't be carried out in humans as experimental animals; common laboratory animal models are necessary. Ehrlich, and others, could do this for trypanosomes because in 1902 Alphonse Laveran, the 1880 discoverer of the malaria parasite, and his colleague F. Mesnil, were able to infect rats and mice with a variety of trypanosome species. The fulminating infections invariably killed, making them suitable for drug discovery. As experimental starters, Laveran and Mesnil demonstrated the curative properties of sodium arsenite.

In 1903, however, Ehrlich remained fixed to his dyes and was injecting them into mice infected with *Trypanosoma equinum*, a biting fly-transmitted species infecting South American horses. He and his Japanese partner, K. Shiga, tried some 500 dyes without success until they injected the azo dye, trypan red. Trypan red cured the infected mice but, like methylene blue and malaria, not in the desired one-shot *therapia sterilans magna* manner. Still, trypan red was Ehrlich's first, true chemotherapeutic success against the trypanosome. Nor was it the red dye's end; over the years, other workers kept tinkering with the molecule and finally a powerful descendent, suramin (Bayer 205) was synthesized. Suramin does not pass the blood-brain barrier, so it can only be used to treat early stage trypanosomiasis. Nevertheless, it remains as one of the sheet-anchor drugs to treat sleeping sickness patients. After trypan red, Ehrlich looked for a better class of drug, and arsenic compounds such as Atoxyl caught his attention.

Simple salts of arsenic were easy to make but were far too toxic to be acceptable as a practical treatment for sleeping sickness. As early as 1858, chemists had learned how to attach nitrogen atoms to carbon atoms. It was called the diazo reaction and allowed the construction of an almost infinite variety of complex nitrogen-carbon compounds. In

1902, the German F. Blumenthal, linked arsenic to a benzene ring, forming a substance he named Atoxyl. Within a year of its synthesis, the German medical profession had adopted Atoxyl as both a tonic to be drunk and to be applied topically for skin diseases. Ehrlich at first dismissed Atoxyl for testing against trypanosomes in mice; he was not about to dignify a tonic cum skin medicine with his lofty standard of experimentation. His attitude was to change in 1905 when H. W. Thomas of the Liverpool School of Tropical Medicine reported that Atoxyl cured trypanosome-infected mice. That same year the British and the Germans began treating sleeping sickness patients in their colonies with Atoxyl. Atoxyl saved lives but exacted a price. As many as 10 percent of those treated became blind.* Moreover, Atoxyl was only effective against the trypanosomes in the blood. It was ineffective against late-stage sleeping sickness with central nervous system involvement. And alarmingly, by 1907, some strains had become arsenic resistant. Drug resistance and toxicity have plagued arsenic trypanocides since that time.

Ehrlich, now aware of arsenic's potential, proceeded in his customary methodical way, consuming thousands of mice and cigars in the process. By 1909, he had tested 591 arsenic compounds without any acceptable success. He had rung almost all analog variations when he synthesized dioxy-diamino-arsenobenzol-dihydro-chloride, a preparation he

* The man who inspired many of the youths of my generation, as kind of remote mentor, to pursue microbiology as an adult career was the bacteriologist Paul de Kruif. In his book *Microbe Hunters*, published in 1926, the lives and works of the scientists he wrote about were heroic. I was able to see the first edition held by the National Library of Medicine in Bethesda, Maryland, and was dismayed to find that the author-hero of my youth had feet of racist clay. Or just as disturbing, it reflected the callous racism of that era. He wrote, in his chapter on Ehrlich, of Atoxyl's blinding toxicity, "Atoxyl had been tried on those poor darkies down in Africa."

was to name Salvarsan. This relatively nontoxic compound, basically arsenic atoms attached to two conjoined benzene rings, cured trypanosome-infected mice as it was shortly thereafter to cure trypanosome-infected humans. But again, only at the early stage of infection. Not even Salvarsan could breach the blood-brain barrier. Still, it was Dr. Phantansus's, as his enemies called him, greatest chemotherapeutic triumph. To his critics, he wryly remarked, "I've had ten years of undeserved hard lines and now all at once I've got a piece of undeserved good luck."

The Salvarsan story has another chapter. Its greatest application would be for syphilis, not trypanosomiasis. Fritz Schaudinn, a renowned protozoologist of that time, and unfortunately a leading authority on wayward garden paths, had in 1905 discovered *Treponema pallidum*, the corkscrew-looking microbe responsible for syphilis. Somehow, Schaudinn believed that it was a protozoan parasite related to the trypanosomes.* This led Ehrlich to the logic that what was good for sleeping sickness might be good for syphilis. There was no practical, effective drug for syphilis, and Salvarsan was an obvious candidate to be screened. Unlike the trypanosomes, however, *Treponema pallidum* has obstinately refused to infect experimental rats and mice; it also could

* This was almost a harmless confusion compared to Schaudinn's most outstanding parasitological gaffe. In 1903, he claimed to have observed the "theatrical entry of a malaria sporozoite" into the red blood cells, as R. K. Knowles commented in his 1928 lectures at the Calcutta School of Tropical Medicine. Schaudinn's authority was so powerful that it overrode all the failures to substantiate his finding and it was to take another forty-five years before H. E. Shortt and P. C. C. Garnham demonstrated that the sporozoite, the infective form in the mosquito vector, went first into a liver cell, where it divided prodigiously (exoerythrocytic cycle) before invading the red blood cells (erythrocytic cycle). It is the exoerythrocytic stage that is responsible for relapses in mammalian malarias.

not be maintained in test tube culture. When Salvarsan was being tested against trypanosomes, Ehrlich had another Japanese scientist, Sahachiro Hata, working in his laboratory. Hata had made the seminal discovery that *Treponema pallidum* would infect the testicles of rabbits and cause ulcerating lesions. Using this rabbit syphilis model, Ehrlich quickly proved Salvarsan's potential to combat the venereal disease that was the scourge of Europe since Columbus may (or may not) have brought it back from the New World. The scrotal scabs of the treated syphilitic rabbits were hardly dried when Ehrlich called on his physician friend, Konrad Alt, to try the preparation in Alt's syphilitic patients. A miraculous winner—Salvarsan proved as effective in humans as it was in rabbits. Celebrating the discovery, Ehrlich's collaborating pharmaceutical firm, Fabwerke-Hoechst, sent without charge, 65,000 doses to physicians throughout the world. The salvation arsenical remained the foundation for the treatment of syphilis until the advent of antibiotics.

Despite their toxicity, arsenicals seemed to be the best pharmacological bet against trypanosomes. New formulations were synthesized and this line of research paid off in 1919 when the Americans W. A. Jacobs and M. Heidelberger developed tryparsamide (for the chemically minded, sodium *p*-glycineamidophenylarsonate). Tryparsamide was unique among the arsenicals in its ability to cross the blood-brain barrier and enter the extravascular compartment of the brain and cerebro-spinal fluid. Late-stage sleeping sickness could now be cured, and countless thousands of Africans were saved. But tryparasamide and the later, related, melarsoprol were, for all their effectiveness, still arsenicals; they were toxic and took their toll. It has been estimated that they have killed or blinded about 10 percent of those treated—an accepted trade-off against a certain death. Attempts were made to detoxify tryparsamide by combining them with neutralizing molecules, such as British anti-Lewisite which had been designed to counteract warfare's poison gas attacks. These formulations partially took the toxic bite out of the arsenicals but unfortunately lessened their ability to cross the blood-brain barrier.

That half-century, from 1900 to 1950, of intense research had yielded a therapeutic armamentarium against trypanosome infections in both humans and domestic animals. The drugs were not the *therapia magna sterilans* of Ehrlich's ideal but they were good enough for colonial government work—and it was drugs or nothing. Vaccines didn't work because of the trypanosome's antigenic agility. All efforts to reduce tsetse fly numbers to the point where there was little or no transmission had failed. *Trypanosoma rhodesiense* was maintained as a reservoir in wild game. There were projects to kill every last antelope; but even this horrific, desperate measure didn't control Rhodesian sleeping sickness. Sleeping sickness continued to be a major health problem throughout British, French, and Portugese sub-Saharan Africa.

Say what you will about pernicious colonialism, the British, French and, at an early time, the Germans and Portugese, did not callously neglect the acute health problems in their dependent colonies. Their pharmaceutical industries, schools of tropical medicine, and infectious disease institutes had expended a great amount of money, time, and talent developing drugs for sleeping sickness and other tropical diseases.* The governments recruited and trained dedicated physicians and scientists who made a lifetime commitment to colonial service. After World War II, the cash-strapped British created well-equipped, well-staffed sleeping sickness research institutes in east and west Africa. British and French sleeping sickness services carried out continuous surveillance and could be quickly mobilized when the outbreak alarms went off. Their large-scale administration of tryparsamide, suramin, and pentamidine dampened endemicity and stemmed epidemics. It wasn't perfect, but thousands of lives were saved.

Then in the late 1950s and early 1960s, the Union Jacks and Tricolors were lowered from the African flagpoles and new national flags raised in their place. With those joyous ceremonies, sleeping sickness services vanished.

I have a good friend at a school of public health who sits in a basement office not much larger than a broom closet—a small broom closet. For many years, he has been, and still is, a respected expert on the biology and epidemiology of trypanosomiasis. When the trypanosome was popular, he was. His research was richly funded, he had his own department, a large office, a large laboratory, coworkers, graduate students, the journals of his choice in the university's medical library—all the perquisites given by administrators to a faculty member bringing in the research bucks (with 40 to

* In 1951, we didn't have a large budget for journals at the West African Institute for Sleeping Sickness Research in Nigeria, but the *British Journal of Pharmacology* was an essential subscription because it published many of the papers on new drugs for tropical diseases that were then being developed. As a kind of exercise to see what was happening in tropical therapeutics, I counted the papers on these drugs over the years: 1951, 69 papers, 12 on tropical medicine drugs; 1982, 183 papers, 0 on tropical medicine drugs; 1999, 509 papers, 0 on tropical medicine drugs.

60 percent overhead going to the administration). Then the trypanosome disappeared from the microbiological cosmos. Malaria became the funders' darling. My friend, loyal to his pathogen lost it all—department, journals, office space—the whole academic nine yards. There is a personal irony in my friend's misfortune. When I started out in 1951, my professor, a famous malariologist, said that malaria was finished, as a disease and a career; he advised that I concentrate on the trypanosome. Now, fifty years later, I can only advise my friend that the trypanosome is dead as a career and he should switch to the malaria parasite. But is the trypanosome truly dead or merely forgotten? The tragedy, for both victim and researcher, is that trypanosomes and their consorts the tsetse flies are alive and well and killing Africans with unprecedented ferocity.

Beginning in the 1930s, every five years, interrupted only by World War II, the global community of tropical medicine experts—clinicians, researchers, and public health workers—assemble at an International Congress of Tropical Medicine and Malaria. My first congress was in 1958, held in Lisbon. At that time, I was on the staff of the West African Institute of Trypanosomiasis Research in Nigeria. My time at the meetings was fully occupied by attending the many sessions and symposia on trypanosomiasis—drugs, epidemiology, entomology, pathophysiology. Almost forty years later, 1998, I was attending my last congress, in Nagasaki. This time I was in the guise of a malariologist co-organizer of a symposium on noncerebral malaria, but I thought I'd sneak over to the other sessions to see what my old friend the trypanosome was now up to. What sessions? For the whole congress there were only four papers presented on the African trypanosomes. Three of those were on "test tube" research carried out in Japan; the other paper, from Kenya, was on a serological diagnostic technique. Was African trypanosomiasis now so negligible that it wasn't worth any significant consideration at the most major tropical medicine meeting? Maybe my trypansomologist friend *deserved* to sit in his broom closet office. Then on the last day of the congress, at what seemed to be an ad hoc session, a doctor from Médicins sans Frontières (Doctors Without Borders) told us how he and other workers of his organization had been desperately battling a sleeping sickness outbreak on the northern Uganda-southern Sudan border. He told of infection rates, neurological involvement rates, and death rates that stunned the few old trypanosomiasis hands present who had experienced yesteryear's outbreaks in other regions of tropical Africa.

With history repeating itself (or continuing), the epicenter of the epidemic has been in Uganda's West Nile District, including the village of Omogo. It was the West Nile Story retold after an intermission of sixty years, when that feverish lady from Omogo, the "mother" of the West Nile virus, was caught in the surveillance net during a sleeping sickness epidemic. Now, however, a new ingredient had been added to the epidemiological cauldron—the refugees of war. For almost twenty years, the Taliban-like government in Khartoum has been waging a grim jihad against their citizens of south Sudan—the Christian and animist unbelievers. There has been an estimated 2 million deaths, compounded by the notorious abduction into slavery of the southerners by the Moslem northerners. Farming and food allocation from western charitable organizations have broken down. The dislocated, starving Sudanese have fled, by the thousands, across the border into Uganda. The Ugandan government maintains that these refugees are infected and responsible for perpetuating the epidemic. Uganda does admit, however, that tsetse fly numbers in the West Nile District are at an all-time high. It is the Gambian form of sleeping sickness, with many of the infected deteriorated to central nervous system involvement. The selfless doctors of Médicins sans Frontières have been at this frontline since 1987. They are in desperate need of trypanocidal drugs, particularly a drug to treat the late stage of the disease. The arsenicals, such as tryparsamide and melarsaprol are especially toxic—blinding and killing these debilitated people. And, at any rate, many strains of trypanosomes are now arsenic-resistant. What these doctors most desire is a supply of the modern miracle drug, the "wake-up-from-the-dead" drug, DL-α-difluoromethylornithine (eflornithine hydrochloride; DFMO). But this has been a money-losing drug, a failed anticancer drug, dropped from manufacture and only recently resurrected to treat hirsute lady readers of *Gourmet* and *Bon Appétit.*

It's a puzzlement. From cancer to facial hair to trypanosomes, the story of DFMO winds through the multinational land of pharmaceuticals. In 1982, Philippe Bey of the Strasbourg, France, pharmaceutical firm, Merrell Tourade et Compagnie, filed as inventor for three patents with the U.S. Patent Office. The first patent covered the synthesis of DFMO. The second patent, based on its effect in pregnant rodents, was for its use as a drug to induce abortion. The third patent was for its use as a drug to reduce the size of an enlarged prostate gland (benign prostatic hypertrophy), a common urinary discomfort in the elderly. As far as one can make out, DFMO came to no commercial use for either the unwanted pregnancy or the unwanted big prostate, but these things have a way of returning to life.

The next great expectation for DFMO was as an anticancer drug based on its action against polyamine metabolism. The cells of mammals and many other animals contain a class of amino acid-like chemicals known as polyamines, colorfully named, for example, putrescine and spermine. Their role in cell biology is not yet absolutely certain but the persuasive evidence is that they bind to DNA and are crucial in cell division and growth. In 1971, D. H. Russell reported in the journal *Nature* that tumor cells have abnormally high concentrations of polyamines. Moreover, it was found that cancer patients have polyamines in their urine which disappear after successful treatment. In 1980, the Dow Chemical Company took over Merrell Tourade. A Merrell Dow Research Institute, the research arm of the Merrell Dow pharmaceutical company, was created with laboratories in Strasbourg and Cincinnati, Ohio. Presumably, the patent for the synthesis of DFMO came as part of the company amalgamation package. Two scientists of the Merrell Dow Institute, Paul Schechter in Cincinnati and Albert Sjoerdsma in Strasbourg, recognized DFMO to be a specific inhibitor of the polyamine ornithine's enzymatic conversion to other, essential, polyamines. Thus, DFMO was considered to have possible anticancer activity by denying the tumor cells' excessive need for polyamines. In experimental cancerous mice, DFMO gave impressive results, but in real-life cancerous humans, it was disappointing and abandoned for commercial consideration.*

DFMO had been great in mice—as an abortion inducer,

* DFMO has not been totally abandoned as a possible anticancer drug. In the current literature, there are numerous papers on the activity of DFMO and its analogs against cancers in experimental animals.

prostate reducer, and tumor killer—but it was a luckless drug in humans. The Merrell Dow scientists, led by Schecter and Sjoerdsma, to their great credit, didn't give up on it and, while seeking other possible applications, came to the trypanosome. A year earlier, in 1979, Cyrus Bacchi and his colleagues of the Haskins Laboratories and New York's Pace University, reported that the African trypanosomes had an ornithine pathway polyamine metabolism that exceeded even that of cancer cells. DFMO, the specific inhibitor of the ornithine enzymatic conversion to essential polyamines was, therefore, logically a trypanosome killer.

Cyrus Bacchi tells of attending a 1979 Gordon Conference meeting on polyamines where he met Peter McCann, a cell biologist at Merrell Dow. Bacchi had not met McCann before, but he "collared" him for lunch, explained his trypanosome-polyamine findings and his belief that DFMO would have trypanocidal activity—and that he would dearly love to get his hands on some DFMO. The next week, McCann sent Bacchi twenty-five grams of DFMO. That Merrell Dow's McCann facilitated the test was a gutsy thing to do and was certainly contrary to the prevailing corporate ideology. By 1980, the pharmaceutical industry realized that natives earning $100 a year weren't going to pay for a new drug's development and licencing costs, let alone earning it a profit. The major firms dropped, even prohibited, screening their new compounds against tropical, mainly parasitic, pathogens. Bacchi tried it out on *Trypanosoma brucei brucei* in rodents, a strain he got from Bill Trager of Rockefeller University. I'm not sure of this, but I think Bill may have gotten it from me when he was a visiting scientist in my laboratory in Nigeria in 1956—the big, multidecade loop of tropical medicine research. Twenty-two years later, there is still a sense of excitement when Bacchi tells of how astoundingly effective DFMO was in his experimental animals. Some animals were so terminal as to be considered as models for late-stage sleeping sickness. The results were incredibly good, even the near-death mice were cured. Later, a consortium led by Schechter, Sjoerdsma, and Peter McCann of Merrell Dow, Bacchi of Haskins-Pace, and Allen B. Clarkson Jr. of New York University confirmed DMFO's powerful trypanocidal effect in experimental animals.

Sjoerdsma's daughter, Ann, has written an article on DFMO for the *Baltimore Sun* in which she recounts how her father and Schechter were determined to get the drug into the hands of a physician for a trial against human sleeping sickness. In 1982, within a year of the rodent experiments, DFMO was delivered to a Belgian doctor, Simon Van Nieuwenhove, who was combating the sleeping sickness epidemic in southern Sudan. The causative strain of *Trypanosoma brucei gambiense* was arsenic (melarsoprol) resistant and he was desperate to obtain a new drug to treat his dying patients. Schecter learned of his plight and began a drug-worthy smuggling operation to get the drug to him in the Sudan. Van Nieuwenhove quickly injected DFMO into two patients with late-stage Gambian sleeping sickness who were consigned to death after not responding to two full courses of melarsoprol. DFMO, diluted in the only available sterile solution—canned fruit juice—brought them back to life. It was a cure that seemed miraculous. The next year, a Belgian patient, comatose with late-stage sleeping sickness, was also "resurrected" by DFMO. Three years later, 1985, the WHO awoke from their own comatose state and arranged for a more extensive trial.

A case report by A. Petru, P. H. Azimi, S. K. Cummins, and

A. Sjoerdsma at the Children's Hospital, Oakland, California, illustrates the dire nature of Gambian sleeping sickness and the curative power of DFMO (*American Journal of Diseases of Childhood*, February 1988). Here is a quote from the abstract of their paper:

> A polyamine biosynthesis inhibitor, eflornithine, was used to treat a 3½-year-old child with newly diagnosed severe trypanosomiasis that had been acquired more than two years previously in Zaire or the Congo. Treatment consisted of 300 to 400 mg/kg/d of eflornithine by continuous intravenous infusion for 25 days followed by 300 mg/kg/d of eflornithine by mouth divided in four equal doses for 17 days. The child's recovery was dramatic, with eradication of blood and cerebrospinal fluid parasites in the first week. Cerebrospinal fluid pleocytosis (*presence of white blood cells*) resolved completely. Her generalized adenopathy and fever gradually resolved. Severe ataxia, inability to walk or to change posture on her own, marked language regression, and lethargy all improved during and after her therapy. Eflornithine was a safe and effective agent for trypanosomiasis with central nervous system involvement in this child.

Thousands of sleeping sickness victims have been returned to life by DFMO. While the pharmacologists may refer to it as DFMO or eflornithine, the Africans with their insightful, accurate vernacular call it the "wake-up-from-the-dead drug." However, this chemical resurrection is neither easy nor cheap, and it certainly is by no means the *therapia magna sterilans* magic bullet. The standard therapeutic course requires daily intravenous injections for at least fourteen days, albeit with relatively little in the way of adverse side affects. It is obviously not a home remedy; DFMO therapy needs careful medical supervision, preferably in a hospital—a luxury rarely available in sleeping sickness foci such as the Uganda–Sudan border area. Nor is it affordable; a course of treatment is estimated to be $600 (the therapeutic course cost has been given variously as anything from $300 to $700). Only through the charity of nongovernmental organizations, such as Médicins sans Frontières, has DFMO been available to the poverty-stricken Africans. Then, in 1999, neither love nor money nor the largesse of charity could obtain DFMO.

Corporate mergers are, to me at least, somewhat murky, but in 1999, Hoechst of Germany and Rhône-Poulenc of France coalesced to form a massive biotech company renamed Aventis. American subsidiaries were established to manufacture pharmaceuticals as well as genetically modified agricultural products such as StarLink corn. Somewhere in the merger agreements, Aventis was given the basic patents for DFMO, now produced exclusively by an American subsidiary. In 1999, not long after Aventis's creation, there was a corporate decision to stop DFMO's manufacture. Nothing about DFMO made money and it was a costly nuisance, corroding the equipment used to make it. However, the original synthesizer, Philippe Bey, says that it is "easy, not difficult" to make. The reaction of tropical medicine's international community was as would be expected. The multinational drug industry was (again) accused of callously ignoring the health needs of the third world. At the 2000 meeting of the American Society of Tropical Medicine and Hygiene, a doctor of the Médicins sans Frontières just back from the Uganda-Sudan sleeping sickness frontier sadly announced that they were down to their last 1,000 doses of DFMO. The remaining stock might last for another six months. After that, the cases that couldn't be treated with pentamidine or suramin at the early stage of infection confined to the blood, would surely perish. This was not a problem that had the dimension or publicity of the South African AIDS epidemic and it was unlikely that some secondary or pirate company would produce the drug.

In early 2001, we were astonished by slick, four-page, heavy paper, colored advertisements in *Gourmet* and *Bon Appétit*, proclaiming the benefit of eflornithine as a salve formulation called Vaniqa. There was Dr. Deborah S. Sarnoff of Cosmetique Dermatology, telling three gorgeous women that Vaniqa would rid them of their uglifying facial hair. The ad announced a contest in which Vaniqaized, now baby-skinned women could send in a photo and essay to win a trip to New York City with all the beautician trimmings. While I did see these ads in the foodie magazines which in post-retirement I peruse, having substituted cooking for laboratory work, I didn't see the Vaniqa ad in *Cosmopolitan.* The *Cosmo* ad I am told, in keeping with the magazine's sexual philosophy, made the pitch that Vaniqa women would be more kissable without their mustache. I don't understand this; I've had a mustache for over fifty years and always considered myself to be eminently kissable.

I can only surmise how DFMO went from trypanocide to depilatory and, as we shall see, back to trypanocide. As part of the general screen, DFMO was tested for its effect on experimentally induced skin cancers of laboratory mice. This, somehow, seems to have led to the entry of another player to find still another patentable application for this versatile compound. On December 22, 1998, David Alberts and Robert Dorr, inventors, were awarded U.S. Patent #5,81,537, "Topical application of alpha-DFMO for preventing skin cancer." The patent was assigned to their employer, Cancer Technologies Inc. of Tucson, Arizona. They bought the DFMO from the manufacturer Marion-Merrell Dow Pharmaceutical Specialties, Inc. (yet another corporate amalgam name change) in Rochester, Minnesota. The DFMO was incorporated into a salve called Vanicream and applied to volunteers with pre-cancerous, sun-damaged skin abnormalities (actinic keratoses). The patent maintains that it reduced these lesions and was deemed to be a skin cancer preventitive, recommending that the Vanicream be applied like tanning lotion before sunbathing. With all that skin smearing of mice and men over some twenty years, it must have been noticed that DFMO removed hair or stopped its growth. This property was ultimately exploited by the corporate trio Bristol-Meyers Squibb, in association with the Gillette Company (in partnership with Westwood-Squibb Cotton Holdings), and they began manufacturing and marketing Vaniqa. There are yet other players; the *Gourmet/Bon Appétit/Cosmo* ads note in fine print that Vaniqa is made under license to two patent holders. One patent dates to 1988 and is held by one Douglas Shander of Gaithersburg, Maryland, no company assignee listed, for "Hair growth modification with ornithine decarboxylase inhibitors." Shander claims that DFMO inhibits, as a topical application, hair growth in "intact, sexually mature males." This would certainly grab the attention of a razor manufacturer like Gillette. The other patent holders being paid off are a mixed bag of inventors: Brian Boxall of Wokingham, England; Geoffrey Amery of Reading, England; and Gurpreet Ahluwalia of Gaithersburg. Their patent, awarded in 1997, looks to me very much like the Shander patent (with the addition of incorporating it into another topical vehicle), but then as I've noted elsewhere, I find the fine workings of the patent process to be totally bewildering.

I have gone through this convoluted DFMO odyssey because I for one became curious how a drug that tropical medicine people considered as "theirs" actually went through all those transformations. What would turn up if we followed other drugs axed from the tropical or orphan infectious disease pharmacopeia? In this brave new multinational world, it is difficult to know who belongs to whom and who really calls the shots. Ultimately though, some corporate power in Germany, France, or the United States decided that marketing DFMO as a prescription vanity cosmetic while denying it to those dying of sleeping sickness did not exactly project a favorable corporate image.

On May 3, 2001, the World Health Organization (WHO) announced in a press release that Aventis Pharma was giving $25 million to the WHO for support of their African Trypanosomiasis program. As part of the "corporate agreement," Bristol-Myers Squibb will arrange for the supply of 60,000 vials of DFMO (presumably purchased with money from the Aventis grant). This is approximately a year's supply of the drug. The WHO will get the drug and, in turn, ship it to the Médicin sans Frontières in Africa. An article in the February 9, 2001, edition of the *New York Times*, "Cosmetic Saves a Cure for Sleeping Sickness," tells much the same story as the WHO press release but portrays the gift horse somewhat differently. As I interpret that article by Donald J. McNeil, the WHO first approached Bristol-Myers Squibb (the maker of Vaniqa) to resume DFMO production for trypanosomiasis. However, it was Aventis (the present holder of the basic DFMO patents) who made the decision to pay up and press for renewed manufacture.

It also seems that the Dow Chemical Company's Akron Manufacturing Inc. in Decatur, Illinois, will manufacture the drug.* The McNeil story indicates that Médicins sans Frontières are not all that happy with these arrangements. Aventis/Bristol-Myers Squibb have guaranteed a single year's supply presumably after which their conscience will be salved and they can continue selling the Vaniqa salve ($54 for a month's supply at your friendly drugstore) without adverse publicity. But Médicins sans Frontières points out that an estimated 300,000 new cases of sleeping sickness occur *each year* in Africa. They need an assured continuing supply of DFMO at an assured affordable price, and say they are able to pay $10 a dose from their budget. It will be interesting to learn whether 2002 or 2003 will be the beginning or the end of DFMO in tropical Africa.

Médicins sans Frontières estimates an annual incidence of sleeping sickness to be 300,000, but really no reliable epidemiologic statistics have been gathered for the past forty years. F. A. Kuzoe of the WHO in Geneva gives a blanket assessment of sleeping sickness occurring in thirty-six African countries south of the Sahara, with 50 million people at risk of acquiring the infection. Kuzoe comments, "Following the attainment of independence from colonial rule in subsequent years, failure by national health authorities to give due attention to sleeping sickness control, due to civil and political unrest, lack of adequate resources and competing national health priorities, has resulted in epidemics and

* Neither my nephew the stock broker nor I (by website search) could find a multinational relationship between these countries. Bristol-Myers merged with Squibb in 1989 and bought the French pharmaceutical UPSA in 1994, but there is no evidence that Aventis or Dow is an arm of the octopus.

the recrudescence of many old foci and the appearance of new ones."

There are 30 million Africans infected with the AIDS virus. In that shadow, the 300,000 annual cases of sleeping sickness may hardly seem worthy of the precious, limited resources. Both the Africans and we of the industrialized nations seem to have been emotionally overwhelmed by the AIDS epidemic, as well as the unrelenting malaria endemicity. We have become almost callous to the continent's other great health needs, such as dealing with sleeping sickness. Perhaps a description of sleeping sickness in one of our own, an American, will help us to relate to the African's plight.

Tourism drives the east African economy. Game parks like the Serengeti and the Tsavo are alive with trypanosome-carrying wild animals, tourists, and blood-thirsty tsetse flies—lots of blood-thirsty tsetse flies. As you can well appreciate, game park tourist trypanosomiasis has not been widely publicized by the African tourist organizations.* The ProMed-mail post (www.promedmail.org) of the International Society for Infectious Diseases is one of the best sources of up-to-date epidemiologic information and case descriptions. Here is a report from that expert tropical medicine clinician, Dr. Claire Panosian of the University of California at Los

* Despite all its other domestic problems, Zimbabwe has successfully controlled tsetse populations using the simple expedient of "artificial cows." These are barrel shaped traps that are baited with a commercial "cow essence" and covered with a residual insecticide. The confused flies are attracted in huge numbers and are killed. Cattle trypanosomiasis, *nagana*, has been almost eliminated. Several years ago my wife and I went on a walking safari vacation in a remote area of Zimbabwe's Hwange game reserve. The artificial cow traps were seen hanging from trees, and I noted very few tsetse coming to attack us or our guide.

Angeles, describing one of her patients who had come to her for a follow-up examination after treatment for sleeping sickness in Nairobi.

> *Seventh Patient Infected with Trypanosomiasis after Visiting Tanzania*
> A 48-year-old American biologist-writer travelled in Africa from December 2000 to March 2001, incurring tsetse fly bites in Zambia and Tanzania. In Tanzania, he was based at a game park near the Serengeti. On Feb. 19, 2001 he developed a fever of 103 (degrees fahrenheit). He presented to Nairobi Hospital where trypanosomes were found in his blood smear and bone marrow aspirate. His cerebrospinal fluid was entirely normal. He was admitted to hospital from 23 February to 5 March 2001, during which time he received 10 days of parenteral pentamidine.

From Italy comes this case:

> A 20-year-old Italian tourist arrived in Kenya on 12 February 2001 where he visited and spent the first night at the East Tsavo Park. On 13 February he went to Tanzania (Serengeti, Ngoro Ngoro). On 22 February he developed a high fever and noticed a big painless chancre (a common early symptom of Rhodesian trypanosomiasis) on the left leg. He was unable to say where he was bitten as there were lots of flies in all the parks. On 25 February he was back in Italy and admitted to hospital on February 26th. Thick and thin films were positive for *T. brucei* (rhodesiense) at high parasitemia, although he was apyrexial on admission. He developed anuria (cessation of urine production) and is still on hemodyalysis but recovering.

Similar case histories have been reported with increasing frequency from throughout Europe. The physician who had treated a Norwegian woman stated, "This is the eighth patient from Serengeti within five months, and the sixth diagnosed in March. We have good reason to assume that we are dealing with a serious problem for travellers to that region *AND* for the local population."

For Claire Panosian's patient, ProMed comments, "This is the seventh case of trypanosomiasis reported in a visitor to eastern Tanzania within the last few months. The lack of comments from the (Tanzanian) Ministry of Health on our reports of trypanosomiasis in October 2000 and these reports may lead one to suspect that there is lack of data on the trypanosomiasis situation in Tanzania. This has implications for the local population, which may experience increased morbidity and mortality from trypanosomiasis without anyone noticing."

"Without anyone noticing"? Has sleeping sickness become an orphan disease in its own continent? Have malaria, malnutrition, tuberculosis, and schistosomiasis become orphan diseases in their own African family of endemic nations? Has AIDS so numbed the thinking of African health authorities and the Western world's AIDS advocates that they have lost sight of reality? Is it tantamount to treason to voice the opinion that no amount of money, no amount of effort will bring a halt to the AIDS plague of sub-Saharan Africa? For this self-inflicted disease there is still neither a vaccine or a one-pill-a-day therapeutic regimen to conduct a campaign. But sleeping sickness, as well as the other diseases degrading the health and dignity of Africans *can*, given the resources, be brought under control. Seventy-five years of productive research in the field and laboratory have provided the methods to do so. Shouldn't we at least consider the possibility that when local health infrastructures are strengthened by accomplishing the do-able, they can then more effectively direct their attention to the great challenge that is AIDS?

Chapter

7

Everybody's Making Money but Tchaikovsky

James (Jock) Williamson and I discovered a formulation, the suramin complexes, that we thought would be *the* answer to African trypanosomiasis (sleeping sickness), especially the economically important form in domestic animals. A single subcutaneous injection of our magic compound had protected experimental cattle against the bite of infected tsetse flies for as long as six months. Our findings appeared in the June 9, 1956, issue of that staid British science journal *Nature*, timed to coincide with the visit of Queen Elizabeth II and her husband-consort to our laboratory in the West African Institute of Trypanosomiasis Research at Kaduna, Nigeria. Jock and I explained to her the nature of the new drug and our first experiments with it—Jock in his rapid, thick Scots–Orkney Island brogue, I in my DNA-coded New York City accent, and our chief technician, Eric Blackie, born and bred within the sound of Bow Bells, making his comments in deep cockney. The queen was later heard to remark that we were a "strange association." There followed a little ceremony in which Jock and I, on behalf of the institute, signed all patent and profit rights over to the Crown. Alas, subsequent, larger studies revealed that in too many animals the drug caused great gobs of skin to slough off. By that time, we were both preparing to move, Jock to the Medical Research Council's Mill Hill laboratories in London and I to the professorship at the university medical school in Singapore. Also, Jock was in one of his periodic anti-English, Scotch nationalist–Stone of Scone moods and dismissed any remedial research with, "It's the bloody queen's drug now, let her bloody well fix it." So we never pursued trying to make the formulation less toxic, although I still think it is basically a good idea for a long-term prophylactic. At any rate, we never had any thought of personal profit; the research was done with the Crown's money and any benefit belonged to the Crown.

Thirty-five years passed before I again had a run-in with the (patent) law. That was the winter of 1996. I had retired from the University of Hawaii the summer before and we moved to North Carolina where our new house, a peculiar Japanoid affair was then under construction by some bewildered Lumbee Indian builders. We too were somewhat bewildered and, I tell you, after thirty years in Hawaii, North Carolina was like being on another planet. To begin with it was *cold,* and the natives spoke an unknown tongue. So, there we were, my wife and I, huddled masses in bed watching a very old movie, when the phone rang and a friendly but serious lady asked if I were Professor Desowitz. She was a lawyer and could she talk to me about *Dirofilaria immitis,* the dog heartworm. Sure, I said; the old movie was boring (as was my new life without parasites), and I was intrigued that the heartworm might be in legal difficulties.

The lawyer said that her firm, based in San Diego, specialized in intellectual property–patent matters and were representing a client named IDEXX based in Maine in a patent infringement dispute. IDEXX was marketing a kit designed to diagnose rapidly and simply a case of heartworm, dirofilariasis, in dogs. It was a serological test, but it was different from the conventional serological test in which the presence of a specific antibody elaborated by the host's immune system in response to an invading pathogen is detected by its reaction to the kit's antigen reagent. The IDEXX kit was based on the fact that the living worm in dirofilariasis excreted metabolic products or shed other protein antigens into the blood circulation. These excretions/antigens had been isolated and, in turn, very specific (monoclonal) and polyclonal antibodies had been produced. These antibodies were the reagents supplied in the kit to detect the circulating free antigens from the worm. It was really a neat idea; in the conventional antibody-detecting serology, the test may be negative if the serum sample is obtained early in the infection when antibody has not yet been formed, or positive late in the infection when the worm is long dead and gone but the antibody persists. On the other hand, if the circulating worm products/antigens were detected, as in the IDEXX method, then it meant there was a current, active infection with living filaria worm parasites.

Now, the University of Missouri's Jewish Hospital and their Gary Weil, M.D., claimed to hold patent rights to the IDEXX test's mechanics and were suing IDEXX for some very big bucks. My immediate reaction to this was what was a nice Jewish Hospital and their physician doing with dog parasites?

The lawyer said that in doing their preparation to defend IDEXX, the literature search found that the first description of the presence of a circulating antigen, the basis for the serological-commercial test, seemed to be a publication by R. S. Desowitz and S. R. Una entitled, "The detection of antibodies in human and animal filariases by counterimmunoelectrophoresis with *Dirolfilaria immitis* antigen," published in the 1972 issue of the *Journal of Helminthology*. It was a study funded by the United Nation's World Health Organization.

The World Health Organization is clearly not in the veterinary research business, but there was a sensible reason for awarding my grant on the posttherapeutic response of heartworm-infected dogs. Humans don't get dog heartworm, although aborted infections causing hypersensitivity and asthma-like pathology have been described. *Dirofilaria immitis* has, however, two filarial, mosquito-transmitted, cousin parasites in humans, *Wuchereria bancrofti* and *Brugia malayi.* The adults of these worms live in the lymphatic ducts and can provoke a chronic inflammation causing fever, edema, and scrotal hydrocele in men. As a worst case scenario, the infection leads to the gross, disfiguring enlargement of limbs and genitalia known as elephantiasis. Millions of people throughout the tropical world (but also once present in early-eighteenth-century Charleston, South Carolina) are infected. There is a very effective treatment in a drug called Hetrazan (diethylcarbamazine), which rapidly clears the blood of the microscopic filarial larvae, the microfilariae, and halt the progress of the disease. The only problem with Hetrazan is that it can be hell on the patient. Severe side reactions to the drug can range from fever to collapse. Let me illustrate the adverse reaction with a story that I've long wanted to fit in somewhere, and maybe this is my last chance.

My friend, the medical anthropologist Dr. Carol Jenkins, discovered a lost Papua New Guinea tribe of hunter-gatherers, the Hagahai. Actually, the Hagahai weren't so much lost as they didn't want to see anybody. After a horrific month-long trek with a police patrol, Carol reached the Hagahai, and they were a medical mess. Their population had fallen to about 300, and the tribe was on its way to extinction. Malaria, often accompanied by a hugely enlarged, fragile spleen, "big spleen disease," upper respiratory infections, and diarrheas were taking their toll. Female infanticide wasn't helping either. Carol had also noticed swollen limbs as well as some cases of frank elephantiasis. After repeated visits, now by helicopter, she became the Good Mother of the Hagahai, bringing in drugs, vaccines, and medical services. She also brought with her the lingua franca of Papua New Guinea, the pidgin English by which the 700 or so tribes, each with their distinct language, communicate with each other and their government.

About two years after Jenkins's long walkabout, I was again in Papua New Guinea. Carol, aware that I had worked on Bancroftian filariasis, including the management of a control program in American Samoa, decided to mount a filarial expedition to the Hagahi. So we assembled our team and supplies from the Papua New Guinea Institute of Medical Research in Goroka, where she was based and I was a visiting researcher, drove to Mount Hagen, collected a young government physician, chartered a helicopter, and made the one-hour flight to the still very wild territory of the Hagahai on the Yuat River. We set up camp in Queen Carol's Castle, a thatched house-hut on pilings with a split bamboo floor that gave a tottering trampoline effect when trod upon. Here, Carol held forth anthropologically and displayed her catholic talents by making great pizza on a propane-fired camp stove.

The first night, about 10 P.M., we turned on the generator, set up microscopes, and assembled the Hagahai to be bled and treated. The night work was necessary because of the remarkable behavior of the microscopic snakelike larvae, the microfilariae whose presence in the blood is a diagnostic criterion of infection. Although they are composed of a relatively small assemblage of cells, the microfilariae "know" to retreat to the deep blood vessels during the day, and with precision, as if each sightless baby worm had a Rolex to consult, flood into the peripheral blood circulation at night to meet their night-biting mosquito vectors. Because of this periodic behavior, it is necessary to draw blood for diagnostic examination between 10 P.M. and 2 A.M.

The Hagahai had reservations about giving blood to strangers. To encourage the rest, the bravest of the Hagahai, the renowned wild pig hunter, Yulein, came forward and volunteered to be first. A blood sample was drawn, a drop placed under the microscope lens and found to be teeming with writhing microfilariae. Yulein was invited to "glass 'em" so that he could see for himself the "snek long rope belong blut," pidgin for the microfilaria (snake) in the veins. Yulein, observing these great-appearing creatures under the microscope was at first astounded, but after thoughtful consideration told us, "me savvy dispela snek" (I know this snake), and then went on to describe a python he frequently saw in the forest. Carol and I were amused by this, in our rather smug civilized way, but upon reflection, we realized its deeper meaning in the way a brain becomes wired to judge the size of perceived objects. Yulein may be only a hunter-gatherer but he is one very smart hunter-gatherer; he can survive in his complex, demanding environment better than could a rocket scientist (or an anthropologist, or a microbiologist). He had never experienced an instrument that would magnify; his cerebral reality was when anything *looked* big, it *was* big.

Yulein was the first to take the snake-killing medicine, Hetrazan pills dosed according to weight. He was told that he might feel a little ill because the medicine was "paitum snek," fighting the snake, and he should go lie down in his hut. Half an hour later, Yulein's son, a boy about seven years old, came rushing in and breathlessly announced, "Yulein dai." Carol, who has a fluent command of pidgin, didn't seem overly concerned. She explained that "dai" isn't really dead, just quite sick. About a half-hour later, the boy reappeared and said that, "Yulein dai tru"—"truly dead." Carol interpreted this as being very sick, and Anian, the institute's man for all problems, was dispatched to administer palliative treatment and comfort to Yulein. By morning, Yulein was bright and fine again, regaling the now treatment-shy Hagahai how his "snek belong blut" and the "marasin" had battled it out. The adverse reaction to Hetrazan is rarely, if ever, fatal—"dai pinis" (die, finish!)—but it is sufficiently severe to produce many refuseniks, and this is a major problem in a filariasis control/eradication campaign, where the strategy is total population drug coverage.

There are no exact experimental models for the adverse reaction of humans to Hetrazan, but dogs infected with heartworm, *Dirofilaria immitis,* come close. The adult *Dirofilaria immitis* lives in the right ventricle of the heart (hence "heartworm") and the pulmonary artery of the lung where it can cause severe, sometimes fatal, disease somewhat similar to congestive heart failure. Like the filarial parasites of humans, the heartworm microfilariae are in the blood where they are picked up by mosquito vectors. As owners of gasping dogs can testify, heartworm is a serious veterinary problem in the United States. Hetrazan is rapidly curative, but virtually all infected dogs react "dai tru" and some even "dai pinis." I proposed to the World Health Organization that my group at the University of Hawaii School of Medicine carry out research on the mechanism of the adverse reaction in the dog heartworm model and screen some compounds that might block or moderate it. The World Health Organization agreed and in 1974 came through with a modest but useful research grant.

I won't burden you with the details of our findings, but one of our opening hypotheses was that the filarial worms were elaborating antigenic metabolic products or shedding such proteins, and/or Hetrazan was causing the sudden, overwhelming release of these antigens. These circulating, soluble parasite products in the serum might trigger the adverse reaction. The next step was to actually demonstrate, by an accepted serological method, the presence of the circulating parasite proteins. Applying a neat, relatively simple technique called countercurrent-immunoelectrophoresis—which, by electric force, drives the antigen and its specific antibody together—we saw the precipitate lines in the gels that gave unmistakable visible proof of circulating soluble parasite antigen. It certainly was not an "incidental" finding, as the Weil/Jewish Hospital patent later described it in their dismissal of others' "prior art." In preparation to publish our findings, the literature search indicated that this would be the first, conclusive demonstration of the filarial soluble antigen in the serum of the host.

There was no consideration of any mercantile profit to us from patenting our discovery. It was work paid for by the World Health Organization and besides, to the despair of my father, an electrical contractor of modest success himself, I didn't have a "business head." In 1976, scientists were not yet legally permitted to profit from discoveries or inventions, emanating from publicly funded research. In the 1940s and 1950s, it seemed that all the songs were syncopated, popularized versions of Tchaikovsky melodies. My antique murky memory is that Jimmy Durante commented on this in a song: "Everybody's Making Money but Tchaikovsky." So we non-industrial scientists were, at that time, poor but proud Tchaikovskys. Later, others sensed the sweet smell of money in the circulating antigen and devised diagnostic tests which were patented as an "invention."

The "inventor" Gary J. Weil and his assignee, the Jewish Hospital of St. Louis, held patent #4,839,275 of June 13, 1989, "Circulating Antigens of Dirofilaria immitis, monoclonal antibodies specific therefor and methods of preparing such antibodies and detecting such antigens." They maintained that the kit IDEXX marketed to diagnose heartworm infections infringed their patent and hired a legal firm to sue IDEXX and get money. However, the lawyer I spoke with explained, in the as yet unfamiliar, exotic language of intellectual property legalese, and in defense of her client, the Desowitz and Una publication could be used as "prior art." Could she discuss this with me? She would come to Pinehurst. Would I read some of the documents and relevant literature? Not yet a consultant, compensation wasn't mentioned, but I was captivated by the prospect of entering a new arena where the contest would not be with the parasite but with the parasitologist.

A week later, a youngish woman bearing a large box of documents, papers, and patents arrived at our little feeder Pinehurst airport. For the next two days, we were to analyze and comment on this material, to begin tracing the origins of patent #275, its shorthand title. Usefully, she had trained in the biomedical sciences before law school. Later, I was to find that most of the staff of her firm—partners, associates, and assistants—had their basic degree in a biomedical discipline. In the former "build-a-better-mousetrap days," patent attorneys usually had an engineering background. Now the action (and legal murkiness) is in biotechnology, and the composition of intellectual property law firms reflect this. As I became more deeply involved, my respect grew for the intelligence, wisdom—and pugnacity—of the lawyers, particularly for my two "handlers" (whom I will call Mary and Lois), responsible for defending IDEXX.

More documents arrived, there were more telephone conversations, and several rolls of fax paper were consumed. Having evidently displayed adequate filarial/immunological knowledge, Mary and Lois decided that I should become an authorized member of the defense, a certified, court-registered, expert consultant. A letter was sent for me to sign declaring that I would be their's alone and not consult with anyone else in this case. Also, there was the query, what was my hourly consultant fee? I hadn't a clue what consultants charge (no business head). During my former life as a working academic, I had frequently served the World Health Organization as a short-term consultant to such places as Fiji, Burma, and Bangladesh. This was not negotiable, and the

World Health Organization decided on terms and conditions, sent airline tickets, and off you went after being cleared by United States security.*

So I called my son-in-law, the secutities lawyer, and asked him what consultants charge. "You're a great, important expert," he dutifully told me, and said I should get the same fee as expert SEC witnesses—$400 an hour. That's astronomical, impossible; no scientist should get that much for just giving advice. So I called Guy Sibilla, my cherished friend in Hawaii, lawyer, adventurer, mountain climber, travel writer. "You're a great, important expert," he affectionately told me, "$300 an hour; that's what physicians get, and you know more than they do," he irrationally added. Still astronomical to my nonbusiness head and I called my legal masters (mistresses?), Mary and Lois, and asked them what I should realistically charge. They suggested $200 an hour and $250 for deposition appearances and testimony in

* There had been a little-known, and certainly little-publicized, agreement between the United States and the UN (when the UN was created) that all Americans, even those serving as short-term consultants for sixty days or less, would have a clearing investigation and government approval before being allowed to take up their assignment. It was a hot time in the Cold War, and the loyalty of all Americans was being put to the test. I guess we were all suspect until proven otherwise, and no flaming Bolsheviks were going to be international servants—even for sixty days. Usually, clearance for the short-term was a short paper affair, having gone through numerous previous clearances. Once, however, the World Health Organization regional office in New Delhi failed to stamp my papers as a sixty-day short-term consultant and the U.S. government acted as if I was going to be a full-time World Health Organization/UN employee. For this, I got the full treatment. FBI agents began to knock on my neighbors' doors, inquiring whether they knew if I had any un-American sentiments; was I a wife beater, a homosexual—or both? Nor did they, nor would they, explain why they were making these inquiries; nor did I know they were making them. As you might imagine, I got some very funny looks from formerly friendly neighbors. Finally, I was made aware of what was happening from a neighbor who was chief of detectives in the Honolulu Police Department who asked me if I was in trouble and could he be of help. A cable to New Delhi cleared the matter.

court. Their instructions gave me some insight as to how lawyers make a living; for telephone conversations, say hello and click the stopwatch; for reviewing documents and writing opinions, click the stopwatch and turn to the first sentence; for traveling to a deposition or conference with the lawyers, click the stopwatch and turn the ignition key. There were other IDEXX consultants, and I am still unaware if we were paid differently. However, during a bit of deposition nastiness, the opposition's lawyer asked me if I knew that consultant X was being paid $250 an hour and what did I think of that unfairness? I think I replied, "best of luck to him" or some other noncommittal nonsense as required by deposition performance.

Having settled the preliminaries, the next step was to go to mini–law school. Basic education in patent law required not only learning terms and language but also a kind of suspension of customary scientific judgement. An objective of the scientific pursuit is to publish, to make known freely the results of research, the interpretation of that research, and the historical bearing of the research of others. Publish or perish is the rule of the academic jungle. Certainly, in the groves of academe one can be the world's most charismatic teacher, but without a solid record of publications, the prizes of tenure and promotion are hard to come by. But I believe the real truth is that we research and we publish, even in our post-tenure, full-professorship years, because it gives intensely pleasurable ego satisfaction, a drive to be the alpha scientist.*

* I give you here the Desowitz Theory of Scientist Arrested Development, which I quote from an introduction in an invited paper commenting on the seminal contribution of electron microscopy to malariology (R. Desowitz, *Journal of Parasitology*, June 2001). "My wife has long been bewildered by what she considers to be the juve-

Publication in a respected peer-reviewed journal requires the successful running of an internal and external gauntlet. First, one most decide that the research findings are sufficiently new, different, unique, or even contrary that publication is warranted. To some extent, this is a subjective decision and comes after a searching review of the relevant published research findings. Then, after writing the paper to the format prescribed by the journal, your paper is submitted. The editor of the journal, usually a distinguished scientist, sends the manuscript to two or three scientists working in the field and in whom the editor has confidence that they will be fair and critical judges. This is the tough phase of the publication trial. Rarely will a referee pass on a paper without comments or demands for change, even when it is recommended for acceptance. Sometimes these are valid requirements, sometimes the referee wants to show that he/she is smarter than the author(s), and sometimes the referee is a competitor working on the same subject who would like to block or delay publication—the race to be first is, or has been, another law of the science jungle. That is the law for "free" science practiced by "free" scientists who had looked down with disdain and pity at their brethren in bondage to the pharmaceutical and biomedical industries, leading lives of frustrating secrecy, necessitated by economics and patent laws—and who make several multiples of their salary. The separation of the two avenues of scientific pursuit

---

nile behavior of scientists she has known—relatives, friends, and, of course, husband. That husband, the psychoparasitologist, has attempted to explain to her that scientists have their emotional development arrested by a lifelong obsession and possession of playthings. Their toys accompany them into the adult estate, and like children they discard the toy of the moment when a new, more amusing one comes along."

was, as we shall see, largely diminished by the legal establishment of the academic/institutional-industrial complex beginning in the 1980s.

Mary and Lois explained that judgments by expert witnesses in patent disputes could not be made in quite the same way as a referee judging a manuscript for a peer-reviewed journal. Scientific expertise had to be accompanied by a knowledge of the special rules of patent law. For this they gave me pages 49–73 of the second edition of *Patent Law and Practice*, published by the Federal Judicial Center in 1995. During the course of my intellectual property education, and in writing this book, I came to realize how little the vast majority of us knows about patent law and how it so manifestly affects our lives. After all, if a single pill that the pharmacist sells to us for $8 can be profitably manufactured by a bootlegger to sell for $.50, then we should be aware of the regulations that permit this kind of discrepancy. Here are the basics as I understand them:

1. What can be patented?
   a. A process. If you combine substance A with substance B in such a specific manner that out comes substance C then that process is patentable.
   b. A machine. If you design a machine to produce that substance C, the machine (in addition to the process) can be patented.
   c. A substance, i.e., a product. Substance C, a mixture of A and B with distinct properties, is a new product and can be patented.
2. When is it patentable?
   The process, product, or machine can be patented if it is:

a. Useful. It's got to do something that advances the technology (in patentese "useful arts") without being for illegal or immoral purposes.
b. New. So what's new? It doesn't have to be a better mousetrap so long as it is a *different* mousetrap. How different? Well, that's the tricky part and the big grist for the arguing lawyers. There is very little that is *entirely* new. Scientists who publish their "new" research findings always cite all the scientists, and their relevant publications, on whose research and methodologies the new findings are based—in patentese, the "prior art." To quote *Patent Law and Practice* (page 52), "Prior art can be an elusive concept because it is not defined in the patent statute nor is there an all inclusive definition in the case law or literature." Therefore, the granting of a patent depends on how convincing the inventor and his/her lawyers are in persuading the patent office examiners that the invention is truly novel.

   Furthermore, it has to be so different that a person having "ordinary skill in the art" would not have come up with the same invention; it would be "nonobvious" in patentese. There is a lot of arguable latitude as to who fits the definition of the person of "ordinary skill"—a creature that has been likened to that other "ghost in the law, the reasonable man." For the '275 patent, I, with a Ph.D. and a D.Sc. and forty-five years working experience in microbiology, could be considered to have the ordinary skills, and I thought '275 was "obvious." But, who knows? On the other hand, a Ph.D. psychologist or an M.D. dermatologist would not have those ordinary skills, despite their intelligence and acquaintance with the processes of scientific discovery.

3. What can't be patented?
   All those great thoughts and theories, no matter how potentially useful, are not patentable. Sorry.
4. What must be in the patent?
   a. Disclosure. The inventor must so clearly and accurately tell all in a written description from which the person skilled in the pertinent art could make the invention and apply it under the described best conditions for its use.
   b. Claims. Here's another rich source for legal wrangling. The patent must state exactly what the invention will do. These are the claims that define the patent's exclusionary power—what others cannot do without infringing (and being sued). For example, if someone made a commercial kit to diagnose human filariasis employing the principles and methodology described in '275's claims for dog filariasis, but with different reagents, would it be a patent infringement?

In May 1996, with my legal tutoring behind me, the '275 patent and background literature read and reread, the history of Dr. Weil's research and career followed, and the outline of IDEXX's defense strategy disclosed to me, it was time to earn the money that had, literally, put the roof over my head. I wrote a declaration which first summarized my career, particularly as it related to filariasis, and a required disclosure, curiously, of how much I was being paid per hour for consultant services. Next came the important part, the reasons why I thought that patent '275 was invalid, why it should never have been issued in the first place. I wrote the declaration in "vulgate science," and it was then translated into legal elegance by Mary and Lois, and sent back to me for signature acknowledging that the document was entirely of my doing. That was to cause stressful moments when I was deposed by an inquisitorial lawyer for the Jewish Hospital who took me to my words, which I hadn't written, and asked me to defend them in patent law terms. Neoconsultants, beware of friendly lawyers bearing strange words.

If I had been asked to referee '275 for a journal, I would have recommended it for publication—with revisions. There were errors and ambiguities, I thought, in the methodology (the Material and Methods section of a journal paper); failure to cite, or not give adequate credit to earlier work (the Introduction section of a journal paper). And, quite frankly, I disagreed with some of the interpretations (the claims) of their experimental results (the Results and the Discussion sections of a journal paper). The Patent Office also seemed to have their reservations and '275 had a difficult passage to legality. A first patent application was submitted to the Commissioner of Patents and Trademarks in December 1983 and "abandoned" without a patent being issued. It was resubmitted in 1985, and the patent examiner not only didn't like the drawings but rejected *all* twenty-one claims. There was a long, thorough critique by the patent examiner, but what seems to have been the kiss of death was her opinion that the foundation of the patent, the isolated *Dirofilaria* circulating antigen, was simply that the stuff didn't make for a patent, since it had the "same characteristics and utility as those found naturally and therefore do not constitute patentable subject matter."

The patent examiner denied the application on other, far-reaching issues. The test's monoclonal antibodies were produced by a kind of experimental crap shoot (not her words). Nothing in the application's written description indicated that the monoclonal could be resurrected by that patent law Everyman, one of ordinary skill in the art. The examiner also thought that a lot of the claimed invention had been done ("anticipated") by others. Then she signed her decision, "Margaret Moskowitz, Examiner, Dec. 31, 1986."

Finality is not in the high-stakes game of patent proceedings. Seven months later, July 1987, Weil's attorney, Donald G. Leavitt, submitted a reasoned argument why the examiner's denial was wrong and why, therefore, a patent should be issued. But Margaret Moskowitz was not to be persuaded and denied the patent application, giving an equally reasoned argument for rejecting the claims of the revised patent application.

The game continued; neither party folded their hand and in May 1988—it had now been five years from the first filing to the last rejection, Leavitt filed for an extension. The government made a little money, some $200 for every filing. In due course, Margaret Moskowitz received another, reworked, patent application, "For circulating antigens of Dirofilaria immitis, monoclonal antibodies specific therefor and methods preparing such antibodies and detecting such antigens" with an amendment arguing for its issuance. It is, in good part, a lawyer's document citing case law and a little perplexing to the scientist who would have to consider such cited precedents as *Uniroyal Inc.* v. *Rudkin-Wiley Corp. Inc.*, or *Strattoflex Inc.* v. *Aeroquip Corp.*

Ms. Moskowitz remained unpersuaded and again rejected the patent application. The disclosed assay may indeed have been an improvement over other, published, assays but it was not convincingly so better as to be patentable. Then she concluded the matter in bold type of capital letters, "THIS ACTION IS MADE FINAL." No is no—every mother's child knows that. But patent lawyers seem to obey another law. Were they rebellious children? On September 14, 1988, the Weil/Jewish Hospital lawyer submitted, "Amendments AFTER FINAL REJECTION." No new experimental evidence, as a reviewer of a scientific paper might request, but reworked arguments why Ms. Moskowitz was in error. I never met or spoke to Margaret Moskowitz, but she must be one tough lady, as she once more rejected the patent. But she weakened, and at this point, only five claims were denied. An agreement was finally reached, and October 24, 1988, was P(atent) Day, when she informed Leavitt that '275 had been allowed. The patent was issued, for a $280 fee, on November 31, 1988. Weil and the Jewish Hospital at long last had their patent and it was a license to sue.

In the United States, heartworm is a major cause for a dogowner to bring his pet to the veterinarian, and a kind of industry has grown around this parasitic nematode. It has been so widespread and pervasive that it prompted the formation of its own organization, the American Heartworm Society, organized in Auburn, Alabama, in 1974. The December 1984 issue of the society's bulletin notes that the circulating *Dirofilaria* antigen described by Weil and his colleagues in the *American Journal of Tropical Medicine and Hygiene* (May 1984) might be the basis for a much-needed, sensitive serological test for the infection. Obviously, there was much profit to be made from a commercial test kit, both for the manufacturer marketing it and the veterinarian applying it. Dogowners might well forgo treatment for their own health problems but would never deny veterinary attention for their cherished pet, no matter the cost.

Around 1983 or 1984, Weil/Jewish Hospital licensed Malinckrodt, a large chemical and biologicals company to produce and sell a kit using their antibodies. From 1985 to 1988, Malinckrodt grossed $455,557 from their "Filarochek" kit sales. How much Weil/Jewish Hospital made from their licensing arrangement is unknown, but if it was 10 percent, it wouldn't make them wealthy but it would certainly be more than pocket change. Nor can this ordinary American not versed in the philosophy and logic of patent law understand how a license can be given while the patent is still in the pending limbo. If the patent fails to be issued, as '275 came so close to that rejection, do you give the money back?

By 1987, other firms had developed and were selling circulating antigen diagnostic kits—"Dirochek" of Synbiotics Corporation in San Diego, "Equate" of Binax in South Portland, Maine, and "Petchek" of IDEXX, also in Portland. The Australians, who also have a big heartworm problem, were also producing circulating antigen test reagents.*

In 1995, IDEXX was sued by Weil/Jewish Hospital for patent infringement. IDEXX was prepared to fight and hired a noted intellectual property legal firm, Lyon and Lyon of Los Angeles and La Jolla, to plead their case before judge and jury. About two years later, each side, having enlisted their expert consultants, made discoveries, taken depositions, and filed motions, the opposing lawyers stood

* With my customary heavy-handed irony, I would note that there were no commercial kits for the immunological diagnosis for the millions of the tropical poor humans with their filariases.

ready to have their day in court. Actually, it was anticipated that it would be a month of days in trial. Mary and Lois asked that I be in Maine for the whole month as a kind of consultant-in-residence. I was semi-retired-free and hadn't had a pristine-fresh lobster for dinner since taking a freighter to Nigeria via Maine in 1956. We on the IDEXX side were confident that our argument was just and that we would win. So my wife and I disinterred our alpine clothing and waited for the airline tickets—which never came.

After a few days of silence I called Mary and Lois and asked what was happening. An evidently disappointed pair of lawyers told me that at the eleventh hour, IDEXX had settled out of court and agreed to pay up and accept a licensing arrangement with Weil/Jewish Hospital. They thanked me for my work, we parted friends and now exchange Christmas cards. Unofficially, by hearsay, I was to learn of why IDEXX folded. The judge in Maine told IDEXX that this was a very complicated case dependent on a jury evaluating a lot of complicated science. Then he allegedly told IDEXX that the jury would undoubtedly be made up of working-class, blue-collar citizens, unschooled in complicated science and susceptible to the cause of a nice hospital wronged by a rapacious corporation. Better to settle for $2 million than to be whacked with a $20 million award. So the story goes. Thus ended the IDEXX affair. Was it worth the telling? Can any broader meaning be inferred from it? Is it a model? Is *Jewish Hospital* v. *IDEXX* a true emulation, a model, of the intellectual property issues that are now of great ethical and political concern in health and biomedicine?

The award of a patent giving sovereign rights for fifteen or twenty years is, to me, a wide departure from the recognition of discovery in science's domain. In the patent process, there is no panel of experts to provide a peer review. There is only a patent examiner who seems to wield extraordinary power. Who is Margaret Moskowitz and why is she able to make decisions that will affect how much a dogowner will pay for a laboratory diagnosis of heartworm? Who are the patent examiners that decide on the availability of anti-cancer drugs, anti-AIDS drugs, and gene products? Are they experts, specialists in the field of the application assigned to them? Is Ms. Moskowitz a parasitologist? Microbiologist? Immunologist? Her name doesn't appear on the scientific society membership lists available to me. Does she, or is she even allowed, to seek expert advice in making her decisions? Why is she immune if there is a subsequent patent dispute as in the case of *Jewish Hospital* v. *IDEXX*? Here is a patent that she repeatedly rejected, even to a FINAL REJECTION, but finally allowed. Why couldn't Mary and Lois grill Margaret Moskowitz in subpoenaed deposition to question her last reversal in favor of Jewish Hospital? I came away from this consultant experience with a citizen's discomfort in the patent process and its disjuncture from the rules of science's game.

There are inconsistencies and misgivings in the patenting of life and life's substances. Only God can make a tree. But could He, would He, patent it? In 1954 someone asked Jonas Salk, "Who held the patent for his polio vaccine?" He replied, "There is no patent; could you patent the sun?"

Chapter

8

The New Guinea Retrovirus and the Federal Bodysnatchers

My first acquaintance with anthropologists was in colonial 1950s Northern Nigeria, and they reminded me of late-Victorian members of the Explorers Club. They lived like natives in small, remote pagan villages (colonial religious taxonomy then being Christian, Moslem, and pagan, i.e., animist) where they studied tribal and clan culture. Their observations were recorded in learned journals and doctoral theses, which have since become Rosetta stones of historical ethology. I met some of these anthropologists in "their" villages, during field trips from the sleeping sickness research institute where I then worked. They, in turn, would come visit me in the luxury of my cool Bauchi Plateau when they needed supplies or, simply, respite; the established academics driving their clapped-out, short-wheelbase Land Rovers and the doctoral students arriving by the "mammy wagon" trucks, which served as a sort of national bus system.

All those I knew were delightful, dedicated, enthusiastic scholars of human society (a society which I found too complicated to contemplate and preferred mindless parasitic protozoa for my subjects of study). They were all, simply, "anthropologists"—not medical anthropologists, or genetic anthropologists, or forensic anthropologists, but simply "anthropologists."

The generic anthropologist began to disappear in the 1950s and 1960s. The biomedical researchers—geneticists, epidemiologists, and microbiologists—discovered the anthropologist's utility in gaining entry into tribal groups as well as their providing a ready-made source of important demographic data. In turn, the anthropologists discovered biology and their calling evolved into specialties. Their influence grew, especially in devising strategies for disease control campaigns such as those against malaria. When the World Health Organization embarked on their Global Eradication of Malaria program in 1952, the then director-general, Brock Chisholm, a Canadian psychiatrist, said that "one cultural anthropologist is worth more than 100 malaria teams." This sentiment did not sit well with the supreme leader of the malaria teams, the Brazilian malariologist Marcolino Candau, who was elected to replace Chisholm and relegate the anthropologist subordinate to the DDT sprayman.

A pioneer in the biomedical-anthropology coalition was James V. Neel, professor of human genetics at the University of Michigan School of Medicine. Neel was interested in the population genetics and disease patterns in "virgin soil" isolated tribal groups, the Yanomamo and Xavante Indians of Amazonia. The Neel team's entrée to the Yanomamo was facilitated by the American anthropologist Napoleon

Chagnon, who had long worked with these people. The introduction of the researchers was followed by the introduction of a devastating measles epidemic that almost wiped out the Yanomamo. In his recent book, *Darkness in El Dorado* (W. W. Norton, 2000), the author, Patrick Tierney, bitterly denounced Chagnon for falsifying Yanomama character and being indirectly responsible for the deaths by measles.* Part I of Tierney's book is titled, "Guns, Germs, and Anthropologists." Thirty years earlier, Neel gave a different take in his summing-up paper, "Lessons From a 'Primitive' People." He wrote: ". . . we have witnessed at first hand the consequences of a measles epidemic among the Yanomama, known from antibody studies to be a 'virgin soil' population with respect to this virus. Although the symptomatic response of the Indian to the disease may be somewhat (but not markedly) greater than our own, much of the well-recognized enhanced morbidity is due to the secondary features of the epidemic—the collapse of village life. . . ."† Thus, Carol Jenkins, when denounced as an exploiter of "her" tribe, was not unique. The new twist in her case, however, was that this was

* Denunciation seems to be an occupational hazard of anthropology. During the Vietnam War, anthropologists were hired, some by the Rand Corporation who had government (CIA?) contracts, to study and be liaison to the Montagnard hill tribes. For this, they were denounced by their professional society and their careers aborted. Yet, those I met during 1972 and 1973, when I led the epidemiology segment of the National Academy of Science's Committee on the Effects of Herbicides in Vietnam, I found to be the most stalwart defenders of the hill tribes. They insisted that the tribal peoples were seriously affected by the herbicides and defoliants, whereas the Committee believed there to be little or no adverse effect.

† In light of the new, "emerging" diseases that have beset industrialized and nonindustrialized populations, Neel was prophetic thirty years ago when he continued, "After witnessing this spectacle, I find it unpleasant to contemplate its possible modern counterpart—when, in some densely populated area, a new pathogen, or an old one such as smallpox or malaria, appears and escapes control, and serious breakdown of local service follows."

an accusation of capitalism at its patent-rights worst in making money from living human creatures—the Hagahai.

Carol Jenkins is a distinguished medical anthropologist. She worked for many years in Papua New Guinea and most recently at the National Institutes of Health where she did research on the behavioral-cultural factors contributing to the epidemiology of AIDS. However, what *really* impressed me when I first met her in 1984 at the Papua New Guinea Institute of Medical Research was that she was the daughter of one of Borah Minevitch's Harmonica Rascals. The harmonica is now near-forgotten, but during my childhood the mouth organ was a virtuoso instrument. The Harmonica Rascals were a wonderful comic act that I occasionally saw at the vaudeville circuit–movie houses in Buffalo and New York City. And to think—I knew and worked with the daughter of a Harmonica Rascal! After a warm-up career as a jazz singer (her husband, Travis, is a professional jazz musician and blows a mean horn), a Ph.D. in anthropology, and field work in Belize, she came to the Institute of Medical Research in Goroka, the Eastern Highlands of Papua New Guinea, where cultural factors in disease epidemiology had long been an integral part of the institute's research. There never had been even a hint of scandal or controversy surrounding this internationally renowned institution. Its director, Dr. Michael Alpers, the bearded, affable former Oxford-rowing-blue, was a circumspect, sensitive leader—as was Carol Jenkins and all the other permanent and visiting staff scientists. The unblemished reputation became clouded when Carol Jenkins was listed as one of the inventors on United States patent #5,397,696, "Papua New Guinea human T-lymphotropic virus." That virus, HTLV-1, came from the blood of a Hagahai.

Center stage of viral pathology is now, rightly, occupied by the Human Immunodeficiency Virus (HIV), the cause of Acquired Immunodeficiency Syndrome (AIDS). However, HIV is only one member of a family of viruses, each with a quirky lifestyle, known as retroviruses. For virtually all life forms—viruses, parasites, plants, primates, protozoa—the canon of creation is DNA (the genetic code genome) to RNA (the assembler) to protein. For reasons best known to themselves, the retroviruses reproduce ass-backward, so to speak, in that they begin with an RNA genome. The retrovirus like most "normal" viruses has a protein-enveloping coat. This protein has a molecular architecture that specifically glues it to the outer membrane of the preferred host cell which is to be invaded. It then insinuates itself into the host cell, where a profound genetic transformation takes place; the virus RNA is converted to DNA, by means of a unique enzyme called reverse transcriptase (from which the designation "retrovirus" is derived). The now naked DNA insinuates itself into the host genome, where like a DNA cuckoo, it directs the host cell to make the viral RNA. The RNA directs the assembly of the virus. That's the way retroviruses are born.

Until 1980, retroviruses were thought to be pathogens exclusively of animals, particularly chickens and cats. Retroviruses killed by turning a host cell cancerous. We had a death in the family when our beloved Burmese, Supercat, was killed by the notorious, highly infectious retrovirus of feline leukemia. Then in 1980, an adult American of African descent died of a very aggressive leukemia, which took the form of a cancerous, uncontrollable production of T-cell lymphocytes (lymphoproliferative). Electron microscopy of the leukemic T-cells revealed virions, virus particles. Several years earlier, a technique had been discovered which enabled the perpetuation, "immortalization," of the T-cell lymphocyte in "test tube" culture. When the virus particles of the leukemic patient were added to the T-cell cultures, they invaded, replicated, and killed the host cells—presumptive evidence of cause and effect. Analysis showed it to be a retrovirus, the first of the family to be discovered as a human pathogen. It was named, Human T-cell Lymphotropic Virus-1, HTLV-1.

Within a year, other African-Americans were identified with incurable HTLV-1 leukemia. Also at that time, a second, incurable, pathogenic guise of HTLV-1 was recognized. This syndrome was variously named Tropical Spastic Paraparesis and Human T-cell Leukemia Type-1 Associated Myelopathy, usually referred to by their shorthand acronyms TSP and ATL. The pathological mechanisms of TSP/ATL have not been fully elucidated, but it seems to be a kind of inflammatory disease affecting the nerves, especially those emerging from the spinal cord at the chest level. The virus-transformed T-cells gather about these nerves and somehow cause them to demyelinate, to lose their sheathing myelin insulation. Paralysis is progressive; death is the outcome.

This was the stuff that infectious disease researchers dream of—a new virus, a new deadly syndrome, and unchartered epidemiological waters. Fortunately, reliable serological and molecular diagnostic/analytical techniques were rapidly devised, and their widespread application revealed that this was a near-cosmopolitan virus, although there were some regional "hot spots" of infection. Retrospective hindsight found the original clinical description of ATL had been made by Japanese physicians who had noticed the malignancy in adults born in the southern islands of Japan. In

1982, Japanese and American virologists began to link those cancers to the newly discovered HTLV-1. Clusters of ATL and TSP patients, serologically positive for HTLV-1, were reported from many parts of the world, with concentrations in the Carribean, the Middle East, southern Europe, and sub-Saharan Africa.

What is the origin of the species? Where did HTLV-1 come from and how did it get into humans? When these questions were asked of its viral cousin, HIV, the answer pointed to a virus in African monkeys and higher apes. Similarly, there is a very closely related Simian T-lymphotropic Virus Type-1 (STLV-1) retrovirus of African monkeys. It shares some 60 to 90 percent of its nucleotide sequence with HTLV-1, a genetic paternity test prompting the theory that STLV-1 is the ancient father of the related human viruses.

HTLV-1 has been a "marker" to trace the earliest human migrations into the Western hemisphere. New World monkeys do not have STLV-1, so at some point in human evolution, the viruses passed to humans in Africa and entered Asia with their migrating hosts. The next dispersal would be with the Asian migration into America. It has been assumed that these would have been Siberians who made that first journey. The medical geneticist James V. Neel found high seropositivity rates of HTLV-1 and a related retrovirus, HTLV-2, in isolated Indian populations of Brazil and Panama.* Neel and his colleagues Robert Biggar and Rem Sukernik have proposed an HTLV-1,2-based hypothesis on the populating of America. They found that while HTLV-2 is common (sero-

* HTLV-2 was recovered from an adult white American with a cancer known as hairy cell leukemia. Whether or not the virus caused the cancer is not known. Since that one case, HTLV-2 has never again been associated with disease, although it shares an 80 percent nucleotide sequence homology with HTLV-1.

logically positive) in Amerind tribes, it is not present in the eastern Siberians from whom they are conventionally believed to have originated. However, HTLV-2 has now been shown to have a high prevalence in Mongolians, which led to the postulate that the American Indians descend from Mongolians/Manchurians who crossed the Bering Strait Land Bridge 22,000 to 30,000 years ago.

The mummy has returned in viral DNA, and it too points to the Asian origins of the Amerinds. Dr. Kazuo Tajima and his colleagues of the Aichi Cancer Center Research Institute in Nagoya, Japan, were able to examine the bone marrow of mummies excavated in northern Chile. In one 1,500-year-old mummy, traces of HTLV-1 DNA sequences similar to present day Andeans and Japanese were detected. They conclude that their finding "provides evidence that HTLV-1 was carried with Ancient Mongoloids to the Andes before the Colonial era." Perhaps it was carried to South America by Dr. Tajima's ancestors. Some five-thousand-year-old Ecuadorian pottery resembles that of Japan's Jomon era (3,000–2,000 B.C.). It is speculated that Kyushu fisherman of that period could have been taken off course by a storm to reach Ecuador's Pacific coast. This has been offered as an explanation of the pre-Colombian introduction of hookworms in tropical America. To that migrant passage, HTLV-1 might also be added.

When larger, comprehensive serological surveys were carried out in many different parts of the world, a curious, still unexplained, finding emerged. The prevalence of infection as evidenced by the presence of antibody was, in some populations, over 15 percent. HTLV-1 had encircled the globe and, even in the United States, a Red Cross cross-sectional serological survey of their blood donors came up with a

0.025 percent positivity rate. R. F. Edlich, J. A. Arnette, and F. M. Williams of the University of Virginia School of Medicine, in their review, state that infection with HTLV-1 is now a global epidemic, with 10 to 20 million people carrying the virus. However, unlike its retrovirus cousin HIV, for which, with rare exception, the serological positive diagnosis is a death sentence in the untreated, only 4 percent to 5 percent of those positive for HTLV-1 will come down with the frank diseases of ATL or TSP. The rest are the silent majority who remain asymptomatic throughout their life. The 5 percent disease rate statistic seems to be pretty well cosmopolitan *except among the Hagahai.* They have a high infection rate, about 14 percent, but none develop overt disease. *All* the Hagahai are asymptomatic, and this fact excited the interest of the retrovirologist researchers at the National Institutes of Health's (NIH) Institute of Neurological Disorders and Stroke. The telling of the connection is a circuitous digression but a connection nevertheless. And it is a good example of chain reactions in biomedical voyages and discoveries.

The NIH's Neurological Institute had a long-standing interest in the neurologic diseases in Papua New Guinea. It began with the discovery of kuru, the first in the family of what is now known as the spongioform encephalopathies, the prion diseases of Mad Cow disease, Creuzfeldt-Jakob disease, and Chronic Wasting disease of Elk. During the 1950s, the Australian administrative and health authorities became aware that in the Fore tribe of the Eastern Highlands, women and children were dying of a strange neurologic disease that they called *kuru,* the shivering death. Neither virus, bacteria, parasite, or nutritional abnormality could be identified as a causative agent. In 1957, the National Institute of Neurologic Disease and Stroke pediatrician, virologist, and nonstop monologist Carleton Gadjusek, along with the Australian researchers Michael Alpers and Vincent Zigas, came to apply his genius to the chaotic dilemma of kuru. Gadjusek shipped the brain of a woman dead of kuru to his laboratory in Bethesda, Maryland, where some of the homogenized brain tissue was inoculated into Georgette the chimpanzee. Nothing much happened, and Carleton being Carleton, Georgette was pretty much a forgotten ape. Then, two years later, the animal attendant noticed Georgette huddled in the corner of her cage, a blank stare on her usually animated face, her body shivering. After a long incubation period, Georgette had fatal kuru. It was transmissible; that is, Georgette's brain tissue caused kuru when inoculated into other animals. Gadjusek believed it was a transmissible virus, a "slow" virus—but no virus could be isolated or seen by electron microscopy. Transmission was by eating infected nervous tissues, and the Fore women and children, but not the men, ate—totally consumed—their dead relatives.

Eventually, the transmissible agent has come to be known as a prion; it has confounded the biologist's comfortable definition of what constitutes life. It is transmissible like a microbe but has no morphological life structure, no nucleic acid, and does not obey life's canon of DNA to RNA to protein. It does not excite an immune response giving rise to antibodies (making laboratory diagnosis difficult). The prion is a rogue protein that somehow makes other rogue proteins in its image (for biology mavens, it is associated with amyloid fibrils polymerized from abnormal monomeric proteins by mutations in lysozyme). Again somehow, they cause holes in the brain, the spongiform encephalopathies. Kuru led to Creutzfeld-Jakob and, eventually, to the deranged cow of Bovine Spongiform Encephalopathy. It also led to the association between Carleton Gadjusek's lab in Bethesda and Michael Alper's institute in Goroka, not many miles from the territory of the Fore. When blood surveys were done in Papua New Guinea, it was customary to tithe a few milliliters to Carleton's crew for diagnostic analyses and to isolate any pathogens, especially those that may be responsible for neurologic disease. In this manner, when Hagahai blood samples were taken by Carol Jenkins's team in 1986 and sent to Bethesda, one of the viruses marked for detection was HTLV-1, the cause of Tropical Spastic Paralysis.

Carol Jenkins came to the Hagahai in 1984, and within a year she began the tribal phlebotomy. Carol treats everyone alike—me, Travis, the husband/jazz musician, Carleton Gadjusek, the Hagahai. She is direct almost to the point of brusqueness, never belittling, and argues from her view that there are universal values without national, tribal, or cultural boundaries. When she persuaded the Hagahai to give of their blood she told them that the procedure would do no harm, that doctors would look for a *binitang*, an insect, in their blood (remember the "snake in the blood" episode; the Hagahai then had no concept of the microbial microscopic), and then appealed to their altruism by telling them that finding the *binitang* would help people all over the world, to make an end, a "liklik shut," to this thing.

Between 1985 and 1988, one hundred and twenty adult Hagahai were bled, the samples temporarily stored in the "Hagahai Hilton's" kerosene-burning fridge and then sent to Goroka and finally air-shipped to Bethesda. Serological tests of this battery of sera showed, as noted earlier, a 14 percent positivity rate for HTLV-1. However, the Hagahai antibodies didn't completely neutralize the standard "cosmopolitan" strains of HTLV-1 from such places as Japan and Africa. This indicated that the Hagahai HTLV-1 was a variant virus. Moreover, it was a benign variant that produced no discernible disease. It was deemed worthy of further, intense research.

The first priorities were to capture the virus and then to continuously propagate it in the culture medium, in vitro culture. In 1989, twenty-five milliliters of blood were withdrawn from each of twenty-four Hagahai adults who were HTLV-1 serologically positive. A helicopter was dispatched from Mount Hagen airport, and the blood samples were delivered to the Institute of Medical Research in Goroka, where they were processed by spinning down, centrifuging, and admixing with a commercial fluid that separated the lymphocytes into a distinct layer in the centrifuge tube. The lymphocytes were collected from this layer, placed in a stabilizing medium, and air-shipped to Bethesda. At the NIH laboratory, a preserving dose of that smelly human/horse lineament, DMSO, was added to the lymphocyte suspension. Now protected, the lymphocytes were put to metabolic sleep in thermos-like canisters containing liquid nitrogen. As needed, the stored cells were reawakened by rapid thawing, placed in a complex sustaining culture fluid to which a dash of Interleukin-2 (IL-2) cytokine had been added.* The IL-2 transformed, "immortalized," the T-cell lymphocytes into a cancerous-like state in which they would divide continuously as long as they were transferred every few days into a fresh culture medium. The transformed Hagahai T-cells were

* Cytokines are secreted by a variety of cells of the immune system and have a great variety of functions. The interleukins are produced by leukocytes and interact on other leukocytes. IL-2 is produced by helper T-cells when stimulated by mitogens or certain antigens. IL-2 does several things, like immortalizing lymphocytes. It also "turns on" cancer-killing lymphocytes and has been used as an experimental treatment for certain cancers (with as yet very modest success).

tested for the presence of HTLV-1 by immunofluorescence, or staining with a specific antibody tagged with a fluorescent dye that lights up when excited by ultraviolet light.

All cultured lymphocytes except one sample from a twenty-year-old male were a bust as in vitro virus producers. HTLV-1 was present in this lymphocyte line, but even after five months of continuous subculture in the Hagahai's now immortalized T-cells, less than 1 percent of the cells became infected. This was not a sufficient yield for analytical work so the research leader, Rick Yanigahara, the Hawaiian-born physician/virologist of the National Institute of Neurologic Disease and Stroke, went shopping for another, more susceptible, T-cell line. His rationale was that patients of ATL have a heavy burden of HTLV-1 in their lymphocytes and, if he could get a line of those cells without the patient's virus, they might also support a more luxurious growth of the Hagahai virus. He found that line, designated MOLT-3, for sale for $205 by the American Type Culture Collection (ATCC) of Rockville, Maryland.* MOLT-3 lymphocytes came from an ATL patient (probably Japanese) and had been deposited by a Dr. Minowada into the ATCC bank some nine years earlier. MOLT-3 and Hagahai HTLV-1 (now designated by Yanigahara as PNG-1) proved, for all practical purposes, to be a biological marriage made in heaven, and within a few weeks their cocultivation was yielding cell infection rates of over 85 percent. With that kind of production, enough virus could be harvested to determine whether PNG-1 was truly a different kind of HTLV-1.

* The ATCC is a semi-public, semi-commercial organization which maintains a "library" of microbial, parasitic, and tissue types deposited by researchers. All are available, pathogen and nonpathogen alike, for purchase by other researchers and nonresearchers.

Modern molecular genetics and modern molecular microbiology has been a wonderfully happy union—especially for the microbiologists. The DNA of the virus's genome, its genetic code, can be dissected apart, amplified, and its characteristic nucleotide sequences portrayed. Also the protein transcribed by the gene can be collected and characterized by sequencing its amino acids. PNG-1 was put through these analytical hoops and compared to "cosmopolitan" HTLV-1 strains, some derived from patients with ATL and TSP. The results confirmed the supposition that PNG-1 was different.

Yanigahara found that PNG-1 was indeed unique; its DNA code varied by at least 10 percent from those of the other HTLV-1 strains—even strains that came from other Melanesian areas, such as the Solomon Islands. If anything, PNG-1 seemed closer to the monkey virus ancestor, STLV-1, than were the "cosmopolitan" HTLV-1 strains. This made Yanigahara wonder about the pedigree of the virus and its human Hagahai host. There is no monkey in Melanesia; never has been. Did the Hagahai migrate from Africa bringing with them the prototype HTLV-1 which persisted as a relic of the prehistoric journey in this isolated tribe? Epidemiological-archeological speculation aside, the important fact was that PNG-1 was different from all other varieties of HTLV-1, and, for Yanigahara and his masters at the National Institutes of Health, it was deemed to be patentable.

PNG-1 is a natural product, a fact of life unchanged over the millennia. It is as unique to the Hagahai as would be an exotic orchid found only in their Yuat River territory. Nevertheless, the legal minds of the National Institutes of Health Office of Technology Transfer considered it to be an invention. It could be patented, its "property rights" given to the employee scientists of the National Institute of Neurologic Disease and Stroke. They would have reigning rights to make money from it. On August 12, 1991, the "inventors" submitted their application to the U.S. Patent & Trademark Office for "Papua New Guinea human T-lymphotropic virus." The inventors listed were the National Institute of Health's Richard Yanigahara, Vivek R. Nerukar, Mark Miller, and Ralph M. Garruto—and, to her ultimate grief and tears, Carol Jenkins of the Papua New Guinea Institute of Medical Research.

Carol Jenkins entered the patent agreement to protect the Hagahai should there be a profit from the "invention." She had called a meeting of the tribal elders and explained to them that she could sign a paper on their behalf and they would get 50 percent of any money if a "liklik shut" or test were developed. Not only did the elders enthusiastically agree but they understood the patent concept. The Hagahai have their own music copyright system whereby individuals or clan groups "own" the songs and dances and must be paid a fee when performed at "singsings" (festivals).

In the patent, the inventors claimed to have invented the Hagahai HTLV-1 variant "in the cell line." It is not clear, but it would seem that the host cell line was the American Type Culture Collection's "store bought" MOLT-3. If so, there would be no Hagahai DNA, as their accusers were later to assert. The inventors also claimed to have invented a bioassay for diagnosis. The technique described had been known for many years and could be performed, as any fool could plainly see, by anyone familiar in the state of the art. This patent was a true and useful invention because, "The establishment of this cell line, the first of its kind from an individual from Papua New Guinea, makes possible the screening of Melanesian populations using a local virus strain." The inventors also stated that the patent would benefit humanity because it "also relates to a vaccine for use in humans to prevent infections with and diseases caused by HTLV-1 and related viruses." Four years later, on March 14, 1995, the U.S. Patent & Trademark Office, with their Einsteinian wisdom, issued patent #5,397,696 to its inventors, who had assigned it to the United States of America as represented by the Department of Health. And that's when the virus hit the fan.

The medical profession has a long association with body parts suppliers. Two hundred years ago, academic doctors had commerce with resurrectionists, bodysnatchers, and grave robbers who disinterred the newly dead for purchase by the anatomists. Not infrequently, when corpses were in short supply, the resurrectionists would respond to market forces by creating merchandise of their own manufacture. The murderers William Hare and his accomplice, William Burke, were two such infamously famous entrepreneurs of the early nineteenth century. Citizen groups of that era rose in righteous indignation, and there was a public outcry to stop the illegal practice. It brought down such eminent anatomists as Dr. Robert Knox.

Fast forward to the late twentieth century. The medical establishment, now in partnership with the burgeoning biotechnology business, was again accused of being in league with bodysnatchers. The new resurrectionists worked on the micromolecular level, taking human genes for sale to the highest bidder. Resentment has been especially bitter when the genetic material has been taken by the biomedical industry of the wealthy world from poor-world people. The governments and intellectuals of the third world have regarded this as colonialism in contemporary form and they have voiced their protests. In the industrialized nations, liberal associations have been formed to protest, on a worldwide forum, what they also view as neocolonialist biopiracy.

One self-appointed guardian of third world genes has been a Canadian group, the Rural Advancement Foundation International (RAFI). RAFI had been monitoring "biopiracy," but there was no one case on which they could focus their full energy for attack. There had been other patent applications involving tribal peoples, including one for an HTLV-1 strain from the Solomon Islands, but these had all been dropped, abandoned, in the face of adverse publicity and the realization that they would probably not be moneymakers. Patent #5,397,696, Papua New Guinea human T-lymphotropic virus, was what RAFI was waiting for. Here was the ideal example of exploitation of the native poorest by those of the wealthiest nation, the Great Satan of Intellectual Property Rights. Even better, one of the inventor-exploiters was an American woman anthropologist—a tool of the genetic capitalists.

In October 1995, RAFI fired their opening shot, a broadside delivered on their Internet website. Their Internet piece was titled "INDIGENOUS PERSON FROM PAPUA NEW GUINEA CLAIMED IN U.S. GOVERNMENT PATENT." It went on to say, "In an unprecedented move, the United States Government has issued itself a patent on a foreign citizen. On March 14, 1995, an indigenous man of the Hagahai people from Papua New Guinea's remote highlands ceased to own his genetic material. While the rest of the world is seeking to protect the knowledge and resources of indigenous people, the National Institutes of Health (NIH) is patenting them." The director of RAFI, Pat Roy Mooney, was then quoted as saying, "This patent is another major step down the road to commodification of life. In the days of colonialism, researchers went after indigenous people's resources and studied their social organizations and customs. But now, in biocolonial times, they are going after the people themselves." Rooney then approached the World Court sitting in The Hague, Holland, and asked them to rule that the patenting of human genetic material is impermissible. The World Court has waffled on this and has not made any definitive pronouncement—not that such a decision could ever be enforced in such genetically laissez-faire countries as the United States.

This attack was too much for Rick Yanigahara, a truly caring person who only wanted to carry on with his research on retroviruses. On the day before the patent was to become legal, he e-mailed Jenkins and said, "Let's drop it, the NGOs [nongovernmental organizations] are driving me crazy." She replied, "The Hags might make some money and I already told them about this. Send the NGOs to me!" Her retrospective comment to me was, "God, what an idiot!"

RAFI got a response more to their liking from the Papua New Guinea government. After meeting with a number of its politician/office holders, the RAFI-ite Jean Christie made the statement to the press that, "The outrageous patent has provoked anger in the Pacific and is a matter of deep concern worldwide." The Papua New Guinea government now professed their anger on behalf of a people that had been of no great concern to them before. There were a few cynics who viewed this as a well-publicized distraction for some *real* biopiracy that was taking place at about that time with government blessings. Malaysian timber interests, having pillaged the forests of southeast Asia, turned to the almost untouched trees of Papua New Guinea. The Malaysians obtained timber concessions from local landowners for very, very little in the way of royalties. In a short time, the stunned clans found that their environment had been largely destroyed. The government interceded, raised royalties from the concessionaires, and took most of the money for their coffers with little of the benefit trickling down to the tribal landowners (Paulias Matane, in the *Independent* newspaper, Port Moresby, March 29, 1996). Now, *that's* biopiracy!

The Hagahai HTLV-1 Affair took an ugly turn in March 1996. Domenic Sengi, the jowly, mustachioed national who has been described, and self-described, as a journalist, student of journalism, and a junior official of the Foreign Affairs Department, was particularly incensed by the RAFI revelations and prodded his superiors to take strong action. Although protests were made to the American ambassador to Papua New Guinea, the only vulnerable warm body that was available for direct attack was Dr. Carol Jenkins.

March 27, 1996: Dr. Carol Jenkins was seated, awaiting takeoff, on a QANTAS flight to take her from Port Moresby, Papua New Guinea, to Sydney, Australia, and then to El Salvador, where she was to participate as rapporteur in a World Health Organization expert meeting on the promotion of sanitation. Before it could move to the runway, the jet was "stormed" by a Foreign Affairs officer and two lawyers. Carol is a big, imposing woman who does not ornament herself with the frills of fashion; she looks like what she is, a working field anthropologist; or to a more fervid imagination, a "green" radical. So when she was forcefully escorted from the plane, the other passengers could well believe that they had had a timely escape from a terrorist. On the tarmac, she was shouted at, vilified as a biopirate, exploiting the innocent Hagahai; her passport was lifted, she was threatened with jail, and finally, she was told not to leave Port Moresby.

Angry and apprehensive, Carol made an appointment to meet with Gabriel Dusava, the foreign secretary. Meanwhile Jenkins's forces were being mobilized; Michael Alpers had been notified, as had the Hagahai. Thus, when she went to her meeting with Dusava, she was accompanied by two Hagahai tribal leaders. One was Yoketan Ibeji; it was his blood that yielded the PNG-1 strain to the National Institute of Health researchers. The other was Korowai Gane who, thanks to Carol, had learned to read and write in English as well as the pidgin lingua franca Tok Pisin. The two Hagahai men presented the foreign secretary with a written statement in which they defended Carol as their protector. She would guarantee that any profits from the patent would go directly to the tribe without it passing through the hands of what they viewed to be a rapacious government.

The first order of the Hagahai was to defend their "Mama bilong Hagahai" whose reputation had been wrongfully sullied. "Carol is a good person and she looks after our interests well. She helps us with many things. She sends us medicine, helped us get a solar refrigerator, and many other things. She is a good person."

Then they showed their remarkable understanding of the patent process. "Now we have heard that the NIH has found a virus in our blood and has made a map of it. The NIH people have also given a paper to Carol Jenkins and told her that in the event that they find this virus and make a vaccine from it, any money that comes will be shared between the NIH and us all. We Hagahai are happy that she signed this paper."

They concluded by declaring that they foresaw a government predation of their patent royalties. "Part of this money does not belong to the PNG government, no way. Why should they get the money? When they get the money they do not think about us, the Hagahai. Also the government did not find this virus, it was Carol and the people in the NIH who found it. The government just think about themselves. So if indeed some money comes from our blood, half will go to the NIH and half will go straight to us, the Hagahai, not to the PNG government. Dr. Carol Jenkins can send it to us. Only Carol knows us. You should not think she will steal; she is a good person."

Em bikpela tingting bilong mipela ol Hagahai
*Tenkyu**

The logic of the facts made Gabriel Dusava reconsider; Carol Jenkins was publicly exonerated and thanked for her many contributions to the health and welfare not only of the Hagahai but all the peoples of Papua New Guinea. But it was a souring experience, and, not long after, she resigned from the Institute of Medical Research and left Papua New Guinea to work in Bangladesh, much to the dismay of the Hagahai. She then joined the National Institutes of Health as an expert on the behavioral and cultural factors contributing to AIDS epidemiology. Even more recently (as forcefully outspoken as ever, "I'm only window dressing; they're not interested in how and why humans get AIDS. Molecules! Molecules! All they know is molecules. The cor-

* This statement translates as, "These are our thoughts, we the Hagahai. Thank you." The Hagahai statement was written in Tok Pisin and the translation by two volunteer community workers, Aaron Petty and Vanessa deKonnick. It appeared in *Cultural Survival Quarterly*, summer 1996, an issue devoted entirely to the Hagahai patent affair.

porate interests are driving the research agenda"), she resigned from the National Institutes of Health, to return to USAID AIDS work in Cambodia.

And what of patent #5,397,696 "Papua New Guinea human T-lymphotropic virus"? Within a year of its being awarded, the National Institutes of Health decided it wasn't worth the furor and decided to drop its claim. Besides, they came late to the realization that the HTLV-1 Hagahai variant wasn't going to be a moneymaker pot-of-gold virus like HIV of AIDS. For one thing, there already were perfectly good commercial diagnostic kits for HTLV-1. In fact, my colleagues and I used two of them (Organon Technika and Cambridge Biotech Corporation) for our rather strange study showing antigen sharing between HTLV-1, including the Hagahai variant, and HIV and the malaria parasite, *Plasmodium falciparum.* And, it was unlikely that pharmaceutical-biomedical companies could be persuaded to invest in a tropical "something nothing" like HTLV-1 when they were reluctant even to invest in an anti-AIDS vaccine. Carol Jenkins objected; she still felt that there would be some royalties that could be handed to the Hagahai for their communal benefit.

In the end, however, the National Institutes of Health abandoned their interest in #5,397,696 and said in effect, "Dr. Jenkins, you want the patent—it's all yours." On May 10, 1996, the American ambassador wrote to Gabriel Dusava that the United States had magnanimously decided to abandon the patent. This could be done, he explained, because "the NIH enjoys new flexibility in this matter thanks to new legislation, the National Transfer and Advancement Act of 1996, which was enacted two months ago." He also said that no one believes there will ever be any commercial benefit, so don't raise the Hagahai's expectations. Then he added that there were a few small details to be resolved—like paying the U.S. Patent & Trademark Office approximately $6,000 from the third party, now the sole inventor to maintain the patent. So, who was the third party? None other than the Hagahai's trustee, Dr. Carol Jenkins. As Carol explains it, with some humor, the kicker in her acceptance was that she personally would have to pay, over the years, that $6,000. "Where am I going to get that kind of money? It was crazy." In the end #5,397,696 was killed, *dai pinis.*

But it was not a small tempest in a small country with an insignificant exotic virus. The Hagahai HTLV-1 Affair was a metaphor, a curtain-raiser for one of the great emerging issues of our time—the taking of life by the powerful from the tropical weak for commercial exploitation, and legitimizing this with patent-intellectual property laws.

When you read written accounts, official letters, statements by the conflictng parties, and newspaper accounts of the Hagahai controversy, you get a sense of both outrage and bewilderment. Science has told the entire world that the miracles of genetic engineering will benefit humankind of all nations, rich and poor alike. Now, when it came to the application of that promise, it was the same familiar story; the rich are to get the drugs, vaccines, and therapeutic genes and the poor are to get more promises, even when they are the sources of those therapies. In addressing the controversy, one concerned group in Papua New Guinea, the Individual and Community Rights Advocacy Forum, was represented by a lawyer named Powers Parkop. Parkop wrote to Michael Alpers that while incredible as it may seem, "U.S. law bars source government and person from claiming rights over such 'property' once it has been biotechnically interfered or developed." Michael replied that this was indeed true but reassured him that the Hagahai's rights and interests would be protected.

Parkop's reading of the U.S. patent law as it applied to human genetic material was confirmed by no less a person than Ronald Brown, the then U.S. secretary of commerce. Brown was able to speak authoritatively on the Hagahai controversy because the National Institute's patent claims on genetic material are pursued worldwide by the National Technical Information Service, an office of the Department of Commerce. Brown said, "Under our laws, subject matter relating to human cells is patentable and there is no provision for considerations relating to the source of the cells that may be the subject of a patent application." The National Institutes of Health, in support of Brown's statement, made their own position clear. Their spokesmen, Amar Bhat of the Office of Technology Transfer, first noted loftily that "The fundamental mission of the National Institutes of Health (NIH) is to conduct and support biomedical research in order to create knowledge that can be used to advance the health of people the world over." Having said that, he added, "NIH has a statutory responsibility to transfer technology to the private sector to ensure that new products are developed for the public health." He explained further, it is the "transfer of technologies to the private sector for the development of new diagnostic and therapeutic agents. . . . NIH must transfer information, materials and intellectual property rights to industrial partners who make the drugs and medical devices. . . ." It is now abundantly evident that this arrangement did not necessarily make the new drugs and devices available "to advance the health of people the world over."

Epidemiologists call it the index case, the first identifiable case that becomes the original source of infection in an epidemic. Was the Hagahai-HTLV-1 controversy the "index case" of a legal patent right exclusivisity war over DNA and other life materials that is to come in the unfolding twenty-first century? The Hagahai may not have been an isolated case of an isolated tribe. A rich vein for genetic material comes from such discreet groups that are, genetically, relatively homogeneous. Their long inbred history usually give them special characteristics that make them valuable to commercial genetic prospectors. This, of course, is also true of the plant and faunal genomic material of isolated ecosystems. It's going to be a battle, sometimes bitter, over the sovereignty of national genes.

The Hagahai case may also indicate who the combatants will be. There were no real villains, and even those with opposite views really had similar objectives. The "radical" organizations such as RAFI were ardent but not fanatic. They, as do so many "ardent" groups, were the ones who called attention to what they saw as a moral injustice. Even Carol Jenkins was sympathetic with their aims. The Papua New Guinea government was acting to protect their citizens from foreign exploitation. I suppose even the National Institutes of Health/U.S. government felt they were acting within their rights. What was good for them would be good for the Hagahai, and all the "Hagahais" of the world. The ethical legitimacy of the government "inventors" may need to be questioned. How did we get to the biomedical-industrial complex run by public—for personal profit—scientists?

The Industrial Strength of the Public Scientist

I am certainly no paragon of morality. I've been accused of wish-fulfillment line calls at tennis—but, hey, none of us are without sin. I like money as well as the next man. I feel that biomedical scientists are inadequately rewarded, considering their contributions. I tell my more affluent physician friends, "Without us guys, you guys would still be applying leeches and drilling holes in heads to let the devils out. Yet, I view with indignant righteousness the practice of biomedical researchers—on the public payroll—being patent inventors, of being in bed, for money, with industry.

I have a cherished friend and colleague of thirty-five years, a distinguished malariologist, and head of a major tropical disease research laboratory. Some years ago—I think it was the time when the HIV dispute with the French had been resolved, and Robert Gallo, then at the National Institutes of Health, was reputed to be collecting about $100,000 a year as his share of the inventor's patent rights—I asked him how this had influenced his own staff's attitudes. He is, like me, of a generation trained in the noncommercial code of science; that the strength of science is free communication. Now, he said, often when he meets with staff working in his laboratory, their first question was not whether their research findings were worthy of publication but whether they were patentable. Then, he added, much of it was unrealistic nonsense; there was little that fit the definition of what was patentable.

The great, if not greatest, gift of biomedical research to the third world would be a malaria vaccine. The Laboratory of Parasitic Disease at the NIH has been a major player in the search for an effective vaccine—indeed, for them, it is the "only game in town." However, even at the NIH, the rule of patent or perish has gained ascendency. A Malaria Vaccine Initiative (MVI) has been created by the NIH to consolidate and fund the research effort. The NIH press release noted, with some pride, that their Office of Technology Transfer was integrated into the MVI at its creation to insure rapid patenting of any alleged vaccine. This is, as any malaria-stricken Hagahai would say in pidgin, *longlong*—crazy. Malaria is a disease of poor people. Any vaccine, to be truly effective, would have to given away. Biopharmaceutical companies aren't going to make (and legally defend) a free vaccine for a billion people unless they adopt the economics of the dressmaker in the Borscht Belt joke who loses money on each garment he sells but makes it up in volume. What then should the MVI, or any other public-supported institution, do if they ever come up with a working malaria vaccine? Perhaps there should be a return to the former rules of conduct. The full nature of the vaccine should be immediately disclosed in a journal publication, establishing it as a "prior art" coming from the public purse. If necessary, they should manufacture it and give it away—even deploy it if no one else will do so. This is what the last truly caring and generous institution, the Rockefeller Institute for Medical Research, did, seventy years ago, after they discovered the yellow fever vaccine.

The Internet has made it impossible for the civil servant to be an anonymous inventor. The computer on my desk is displaying who of the NIH malaria research group has applied for a patent. David Kaslow (now at Merck) has patent #5,217,898, for a transmission-blocking vaccine. Michael Good (now in Australia) has patent #4,886,782, for a malaria immunogenic antigen. John Dame holds #4,707,357, for a vaccine antigen and the genes that make it. Tom McCutchan has two patents, #4,693,994 and #4,707,445, for an immunizing peptide antigen and its encoding gene. Tom Wellems has #5,130,146, for a recombinant clone of a *Plasmodium falciparum* gene expressing an immunizing antigen. And the leader of the group, Louis Miller, no longer quite so noncommercially pure of heart, is the inventor of #5,198,347 and #5,993,827 for *Plasmodium vivax* of humans and *Plasmodium knowlesi* of monkeys, Duffy red cell receptor factors—vaccine possibles. If I've left anyone out, I apologize. All those patents and not a single practical, effective malaria vaccine. *Longlong*!

Military biomedical researchers are also in pursuit of the patent. The most flamboyant appeal to commercialize federal research for fame and fortune that I have seen are the two slick, colorful brochures that the United States Army's Walter Reed Army Institute (WRAIR) distributes to its staff. Titled "Intellectual Property and Patents: Products of the Mind" and "Technology Transfer: Beneficial Collaboration," they instruct the military scientists that if they have research that they consider salable, they should get in touch with Mr. Charles Harris, WRAIR's patent attorney and the director of Research Marketing.

Those scientists are not evil people succumbing to the temptations of Mammon. Some "public inventors" are obeying, as an act of self-preservation, the dictates imposed by their administrators. Others honestly believe that the only realistic way to transfer the products of their research into tangible products of public benefit is by cooperation with industry. Others view it as a right-by-bonus to enrich themselves through their research. But there are others who just say no. Whatever their motive(s), "public inventors" are acting legally; indeed they are encouraged to do so by statutory acts of Congress.

In 1980, the 96th Congress passed the Stevenson-Wylder Technology Innovation Act. The banner summary declared its intent, "To promote the United States technological innovation for the achievement of national economic, environmental, and social goals, and for other purposes." In the Reaganactionary climate of the 1980s, this act returned to the economic philosophy that the business of the United States is business. All federally funded research would now be accessible to industry: "The federal government shall strive to transfer federally owned or originated technology to the private sector." However, it wasn't to be a total giveaway; the marriage of state and industry was to be legitimized by cooperative agreements, partnerships, and limited partnerships. And in one of its most far-reaching provisions, the Stevenson-Wylder Act permitted employees, even former employees, to commercialize inventions they made while in the service of the United States government.

From 1980 to 1995 successive legislation cemented the academic/institutional-industrial complex:

1980. Bayh-Dole Patent and Trademark Amendments Act permitted universities and not-for-profit organizations to patent inventions that had been developed with federal research funds.

1986. The Federal Technology Transfer Act further liberalized technology transfer and cooperative research agreements. It allowed patent royalties to be given to federal employees and federally funded researchers who would get the first $2,000 and 20 percent thereafter, up to $150,000 each year over the twenty-year life of the patent.

1989. The Federal Technology Transfer Act extended the largesse to private companies working under federal contracts.

1995. The Technology Transfer Improvement Act guaranteed industry the right to exclusive license in cooperative agreements.

To the cynics, Stevenson-Wylder et al. may seem to be yet another confirmation that we have the best government money can buy. Another example that the pharmaceutical industry with their lobbyists and massive money contributions have again had their way. This may, in part, be true, but for another perspective we must return to the industrial age of the 1980s when the health of General Motors, in the generic and real sense, was failing. General Motors had manufactured the Corvair, and Toyota the Corolla. If more were to be said, some of the words would be Sony, Honda, Hitachi, Canon—all Japanese "words." In 1980, the Japanese were clearly making better, more innovative products, especially automobiles and electronics, than the Americans—despite the foundation of those products which, as often as not, were American inventions.

The Japanese had, with their government's support, heavily invested in applied research to improve their products and methods of manufacture. The strategy worked; their automobiles, cameras, televisions, and microscopes were a logarithm better than ours. Although we might wryly joke that the Sony telly came from the "country who brought you Pearl Harbor," the Japanese were rubbing our national nose in their industrial superiority. Stevenson-Wylder sought to redress our industrial weakness by encouraging new technology to make Americans competitive with the Japanese. This would be accomplished by giving private enterprise access to the immense variety and depth of federally funded research. And to keep an eye on what our Japanese competitors were up to, the act established a National Technical Information Service "to monitor Japanese technical activities."

There is, however, an arena where America has been the undisputed alpha gladiator—the biomedical business. The Japanese have had the cohesive discipline to fabricate a Lexus and a Walkman with great precision but, essentially, failed in biomedical research—the innovative discoveries that lead to new, superior therapies. It seems to me that first-class biomedical research requires a measure of zaniness, indiscipline, and even eccentricity. The Japanese, with their rigid, Prussian-like academic system, as well as their misogynic failure to foster women's careers in science, simply did not create the nourishing environment for excellence in biomedical research.

Is the Japanese biomedical mediocrity a lesson for America? Will the collusion of academic researchers and industry forfeit the freedom that has allowed our unfettered inquiry and inventiveness? There are immense new challenges now confronting the world. Biomedical science requires the best of its "prepared minds" to save us from those threats—one being global warming.

Chapter

9

We're Having a Heat Wave, a Tropical Heat Wave

For the sake of narrative simplicity, I will use wife to husband, but it is bidirectional in heterosexual marriages. Your wife asks you a question. You answer. It's not the right answer. Time passes. She asks you the same question. You give her the same answer. It's not the right answer. And so it goes. It's like that with global warming and the Republicans. The president asks, "Is there global warming?" You (i.e., the science community) answer, "Yes, there is global warming." It's not the right answer. And so it goes.

Its called a peta, the fifteenth power—a petagram when expressed as grams. The petagrams of carbon inserted into atmosphere each year from the combustion of fossil fuels look like this: 7,000,000,000,000,000.

A good entry source for the global warming dimensions is the Pew Center on Global Climate Change (www.pew

climate.org). They cite the estimates arrived at by the authoritative Intergovernmental Panel on Climate Change. By the year 2100, the average global temperature will be 1.3° to 4.0°C (2.3° to 7.2°F) warmer than today. The best estimate is that the earth will be 3.5°F hotter than it has been for millions of years.

From all the numerous analyses and supercomputer simulations, there is little doubt that global warming is a very real, man-made (anthropogenic) phenomenon. However, profound climate changes, discernible in the paleontological-geological record, have taken place during at least the last several million years. It was sufficiently warm 125,000 years ago that the hippopotamus was happily plodding about in the marshes of balmy Yorkshire, England. Twenty thousand years ago that same Yorkshire swamp was covered with a six-foot-thick layer of ice. During that last ice age, the ice sheet extended into the northern continents, and the sea level was as much as 400 feet lower than it is now.

This background "noise" of natural climate change is caused by small changes in incoming solar energy. The earth's orbit varies, at times becoming more elliptical, which puts it closer or further from the sun. The earth is a spinning top with a varying axis and a wobble. This too changes its orientation to the sun. But orbit, tilt, or wobble notwithstanding, the 30 percent more carbon dioxide that humans have put into their air over the past 150 years has made the average global temperature warmer. So far, it is a modest rise but even so it's had drastic effects causing heat waves, droughts, floods, and even blizzards as more water evaporates from the oceans. If economic evidence is needed, a spokesman for a major reinsurance company complains that their disaster-related payout for 1998 was twice that of the previous years.

He asks the weathermen, "Are we in a time of a permanent El Niño?"

The projection of average global temperature change doesn't give a full picture of what is expected to happen in the near future. Climate change will not be globally uniform; the largest magnitude of warming will likely occur in the northern temperate zones. Here, the heating up will be more pronounced in the night than the day and to add to the discomfort, humidity will also increase. Every three years, spring will arrive one day earlier. There will be environmental changes, demographic changes, agricultural changes, water use changes—changes that are grist for the epidemiological mill. The wisest opinions hold that global warming already has impacted on health, particularly as it is affected by the infectious diseases. One of the disease sentinel canaries testing hotter airs has been the pathogen-bearing mosquitoes, particularly those of malaria. Articles, books, and workshop proceedings have been considering how malaria may be used as the indicator of global warming.

"Great to see you again," "Nice to meet you after all these years," "What do you think this meeting is really all about?"—the usual pre-session exchanges. This casual, customary exchange took place at the elegant Lausanne Palace and Spa in May 2000, where there was a bottle of Pol Roger champagne to accompany the muesli at the breakfast buffet. The fifteen participants of a workshop called "Contextual Determinants of Malaria" had been gathered together by the organizers, Elizabeth Casman and Hadi Dowlatabadi of the Carnegie Mellon University's Department of Engineering and Public Policy (a curious but intriguing academic amalgam). Our assignment was to ponder upon the determinant of global warming on the epidemiology of malaria—that is, its global distribution, intensity, and anopheline vectors. The fifteen ponderers were a mixed bag of malaria experts from India, South America, Europe, Southeast Asia, the United States and, by proxy, Africa. There were orthodox, field-oriented, general purpose malariologists; orthodox, field-oriented medical entomologists; administrator-malariologists from WHO's central casting in Geneva and from regional satellite offices; laboratory research-type, mathematical epidemiology modelers who may never have seen a malaria parasite under the microscope or a malarious patient but had the data and knew the numbers; and climate change experts. There were also three representatives from the workshop's major angel, ExxonMobil. Later, disclosed in the acknowledgments of sponsorship were the American Petroleum Institute, the Electric Power Institute, along with the National Science Foundation and the National Oceanographic and Atmospheric Administration. That evening at the informal reception, the ExxonMobilites were asked why they were paying for the meeting. They explained that malaria was a health problem for the workers in their tropical operations. And then they said candidly and cryptically, "There is of course, the problem of emissions. . . ." Was the workshop a kind of preemptive legal strike, asking the right question and hoping to get the right answer?

Here we were in Switzerland, modern-day malaria prophets attempting, like the ancient oracles, to divine an impending disaster. When the bones were cast, the tarot cards read, and the accepted notions reviewed, the portents were that mosquitoes would flourish in the new warm, humid environments, breeding faster, drinking blood more frequently, and living longer. Malaria, the epitome of a tropical disease, would surely follow. More infections would be due to the dangerous malaria parasite *Plasmodium falciparum* which requires a minimum of 18°C (65°F) to complete its life cycle, whereas the nonlethal *Plasmodium vivax* can make do with a cooler 16°C (61°F). Ergo, as the temperate zones turn tropical, they will become malarious—a killing malaria. The oracle has spoken. Leave your offering at the temple altar.

Then the iconoclast oracles, led by Paul Reiter, the entomologist-historian of the CDC's dengue unit in Puerto Rico, presented a different view that was closer to reality (see his comprehensive review: Climate change and mosquito-borne disease in *Environmental Health Perspectives*, March 2001, 109:141–61). It is true that of the approximately 3,500 species of mosquitoes, including the malaria-transmitting anophelines, most are tropic and subtropic—but not all. Mosquitoes are almost everywhere, adapting to virtually every habitat that is not permanently frozen. And the most massive numbers are not in the Congo but in the sub-Arctic after the spring thaw. Clearly, malaria is *not* a tropical disease. One hundred years ago it was endemic in Sweden to the 15°C (59°F) isotherm. Until 1955, Poland was wracked by severe seasonal outbreaks. Turn of the twentieth-century America was malarious in almost all states east of the Mississippi. DDT, better housing, health care, and water management has eliminated endemic malaria in the United States and Europe.* That is, the parasite has been eliminated. The anopheline vectors remain in place, waiting, waiting. These

* There are several thousand cases of malaria seen by physicians in the United States each year. The great majority are in travelers returning from a malarious region or in immigrants from endemic countries. However, in recent years there have been several bizarre cases of malaria in the New York–New Jersey area in people who have not traveled and so acquired the infection from local anopheline mosquitoes.

potential vectors don't need global warming to pioneer their way to the north, although it is possible that their tropical cousins will join them in the warmed-up America and Europe. Moreover, any mini-outbreak in an industrialized country would be swiftly contained. The Carnegie Mellon meeting concluded with the right answer. ExxonMobil need not worry; no one is going to sue them because they got malaria in Detroit made tropical by the emissions from a Detroit-built SUV.

While global warming has not made Finland or Maine malarious, it has, in fact, brought the disease to areas formerly malaria-free. Malaria now occurs in newly warmed highland areas and semi-desert areas in Africa and New Guinea, where anopheline mosquitoes were always present but the ambient temperature had been too low for the malaria parasite's development (extrinsic cycle). Take the case of Wajir, a town in the semi-desert of northern Kenya, an environment that restricts the breeding of *Anopheles* mosquitoes. In November and December of 1997, the weather changed, propelled by El Niño (which is considered by some experts to be influenced by global warming). Wajir was inundated by torrential rains. Anopheline mosquitoes proliferated, and malaria was soon to follow. Of the 60,000 inhabitants, 40 percent came down with malaria. They had not developed the protective immunity acquired by people exposed to constant malaria exposure; 1,500 died, 108 of which were children under five years of age. Similar examples of "new" malaria can be given for the highland areas of Burundi, Zimbabwe, Madagascar, New Guinea, and Kenya. The consensus was that global warming would have only a "marginal" effect on malaria endemicity. Small niches such as mountain and arid environments could be affected. A cascading effect was considered but deemed unpredictable from the presently known facts. The creation of new semi-deserts in tropical zones could actually reduce malaria. But this would cause population movements and create new, man-made habitats for the malaria mosquitoes. But a Wajir villager whose child has died of malaria would not consider his situation as "marginal."

If malaria is a false canary that doesn't drop dead in the global warming mine, what about other arthropod-borne diseases—those transmitted by ticks or fleas? In these infections, we may have firmer evidence, from areas as temperate as Togliatti-Stavropol in Russia to Sweden's Stockholm county, of climate change's impact on the public's health. Stockholm we know, but Togliatti-Stavropol? Its tick-transmitted outbreak of the infection had the strange name, Crimean-Congo hemorhaggic fever.

Not many under seventy years of age would remember him now, but at mid-twentieth century, Palmiro Togliatti nearly made Italy a Stalinist satellite. Our small cluster of American graduate students in London shortly after World War II would, from time to time, flee Old Blighty to go to Italy for a square, spiced meal and a glimpse of the sun. During one of those visits to Rome in 1949, I witnessed a mammoth hammer-and-sickle parade led by the burly Togliatti demanding a communist takeover of the government. It never happened, but for his service to the world Red order, Stalin changed the name of the southwestern, Volga region city of Stavropol to Togliatti (various atlases also give it as Tolyattigrad or Toljatti. Confusingly, it is not certain if it has reverted back to Stavropol after the breakup of the communist regime). The Stavropol district is a steppe at the foot of the Caucusus, agriculturally rich and rich with meadows and forests. Beginning in 1998, there were unaccustomed outbreaks each spring of Crimean-Congo hemhorrhagic fever, a tick-transmitted virus disease that kills about 25 percent of the infected.

On a warm March day in 1998, Ivan Doervitch of Stavropol/Togliatti went to visit his brother, who had a wheat and sheep farm twenty miles away. Doervitch was hot; he had spent his boyhood on a cooperative farm, but city life had made him soft, and he was sweating after a short walk through the sheep pasture. But it was not all a physical softness, he thought. The seasons had been crazy these last years, warm winters without snow. His brother told him that he may have to give up wheat farming for another crop that would tolerate the higher temperatures. He had read in the newspaper that even Murmansk was warmer, that ships could make the northern passage to America. America again! The newspaper said it was warmer because America was putting all those hot gases into the air. And the Americans complained about a little thing like Chernobyl. He looked down and saw ticks crawling up his pants leg. Damned bugs were everywhere this year, and he brushed them away.

Doervitch returned home to the city, relaxed in a hot bath, and when he dried himself found two ticks, their heads embedded in the flesh near his groin. A burning cigarette applied to their asses took care of those vermin. Then for antiseptic, he splashed some vodka over the bites and for good measure drank a glass of the antiseptic. Five days later, he came down with a severe case of the "flu." He had a fever, chills, a piercing headache, his muscles ached. He was nauseous, vomited, and had diarrhea. When he looked in the mirror, he saw a face flushed borscht red. The next day, he was worse, much worse. His trunk was covered with a rash.

There was profuse bleeding from his nose and gums. This was no flu; he was frightened, and with the help of his neighbor, Doervitch managed to get to the Stavropol/Togliatti Central Hospital. He was admitted to the infectious disease ward, examined, and a blood sample taken. The chief doctor, Olga Balaban, knew the diagnosis even before the serum sample had been analyzed by the Sanitary and Epidemiological laboratory for specific antibody; it was the Crimean-Congo fever. There were thirty-seven patients like Doervitch in her ward. Three days later the laboratory report confirmed her clinical impression. Luckily, Doervitch was alive and recovering. She lost about a quarter of her patients to the disease. Doervitch went home from the hospital later in the week and in his last conversation with Dr. Balaban she explained to him that he had had an untreatable viral disease that was transmitted by ticks, injecting the pathogen when biting. Normally, the virus "lived" harmlessly in domestic animals—sheep, horses, cattle, and goats, as well as in some wild animals like the hedgehog. Probably some birds were also infected; she had read about an outbreak of Crimean-Congo hemorrhagic fever in workers in a South African ostrich abattoir. It was present in many parts of Africa, the Middle East, and central Europe, although it had been a relatively rare infection in humans. However, the warm winters had explosively increased the tick population; a walk in the surrounding woods or fields was sure to bring tick bites.

A year later, Doervitch was still convalescing and unable to work. He was tired all the time. He had become bald and didn't hear very well. Damn those Americans and their hot gases.

Further to the northwest, Sweden has also been experiencing an outbreak of a climate-influenced, tick-transmitted viral infection—the tick-borne encephalitis (TBE). TBE also begins with a walk in the tick-infested bosky wood or meadow. It incubates for about a week after the tick bite and then begins with the flu-like symptoms of fever, chills, headache, and muscle ache. In the ensuing acute phase, the two tick-transmitted viruses depart in their clinical ways. Whereas the Crimean-Congo virus is hemorrhagic, manifested as rashes and bleeding gums and nose, the TBE virus goes for the brain and its covering tissue—it is a meningo-encephalitis. A neurological disease, an inflammation of the brain, the signs and symptoms of TBE are tremor, intense headache, and personality changes. Most cases of TBE recover naturally with few aftereffects, but a few progress to more severe motor derangements that may end fatally. In its epidemiological natural history, the TBE virus is transmitted mainly by the brown, eight-legged "castor bean" tick, *Ixodes ricinus.* Similar to the Crimean-Congo disease, TBE is a zoonosis with natural reservoir hosts in wild and domestic animals—deer, small brush-inhabiting mammals, cattle, and sheep.

In 2001, Drs. Elisabet Lindren of Stockholm University and Rolf Gustafson of Stockholm's Huddinge University Hospital reported that over the past twenty years there has been a rising incidence (cases per year) of TBE in Sweden. In attempting to explain why this once relatively rare infection in humans has become so relatively prevalent, Lindren and Gustafson turned to what has become the usual epidemiological suspect—global warming. The weather data from 1960 to 1998 revealed that Sweden, along with Europe's northern tier, had markedly and progressively warmed since 1980. They wrote, "there has been a global warming trend during the last 2 decades of the 20th century, with the years 1990, 1995 and 1997 having the highest mean temperatures ever registered in the northern hemisphere," and "The spring vegetation season has advanced by 12 days on average between 1960 and the mid-1990s." In these climatic changes, Lindren and Gustafson saw the link to TBE: "The findings indicate that the increase in TBE incidence since the mid-1980s is related to the period's change toward milder winters and early arrival of spring."

The increase of tick-borne infections has been of particular concern to the Swedish military. Their troops train in tick-infested areas where they are at risk to TBE and Lyme disease. The military doctors in their search to find a more benign, natural, and lasting repellent than chemicals such as DEET undertook a trial in which fifty marines were given 1,200 milligrams each day of *Allium savitum* in capsule form. Fifty other marines were given dummy placebo capsules. Each marine kept a "tick diary," in which he recorded the number of tick bites for that day. After ten weeks, the code was broken and the diary data analyzed. While both groups had been bitten by ticks, those who had taken the *Allium sativum* pills had statistically significant fewer bites than the placebo group. *Allium sativum* is garlic.

Earlier, warmer springs and prolonged autumns leading to shorter, warmer winters has led to a burgeoning tick population feeding more frequently. Moreover, the warming of the northlands has allowed the ticks to advance into higher latitudes. The ticks flourished because there was now abundant food for them. Warmer weather leads to more vegetation and more food for the ruminants, especially deer. Well fed, their population expanded. Now everything in the TBE food chain prospered under the warming sun—more ticks that fed on more deer, and the TBE virus enjoying the augmented populations of both their arthropod and mammalian hosts. Meanwhile, the abnormally wet and warm weather was bringing yet another virus to humans—American humans.

Across the Atlantic, in an area called the Four Corners, where the boundaries of Utah, Colorado, New Mexico, and Nevada meet, that same change in the weather was making the desert bloom. One desert's denizen, the deer mouse (*Peromyscus maniculatus*) thrived on the new, abundant food source, and with a full belly, it bred like, well, deer mice. During 1993, it is estimated that their population increased an astounding twentyfold.* It was here in the Four Corners in that year that local residents, mainly Navajo Indians, sickened with the "flu," then became short of breath, then breathless and then, for 40 percent of the afflicted, dead. The disease came to be known as the Hantavirus Pulmonary Syndrome.

The hantavirus family came to American attention during the Korean War when, from 1951 to 1953, over 2,000 American–United Nation troops developed a severe hemorrhagic fever. Transmission experiments in which urine and blood from patients were inoculated into some brave and/or ignorant volunteers proved that the cause was a pathogenic virus. Virologists named it the Hantaan virus for the river in Korea, an epidemic focus. It took another twenty-five years to explain the Hantaan virus's facts of life. It is a virus of rodents—later study on the enzyme-amplified genome indi-

* While primarily a desert animal, the deer mouse is also found throughout the Americas. It ranges through much of Canada in the north to Brazil in the south.

cated that the virus had paired with rats and mice approximately 20 million years ago. It is transmitted to humans via contaminated rodent droppings or urine. It is an RNA virus that the taxonomic (name calling) virologists included in the *Bunyaviridae* family. Serological research showed there to be more than twenty species/varieties of Hantaan virus widely distributed throughout Asia and Africa where they caused hemhorrhagic disease complicated by kidney damage. But until 1993, Americans thought themselves safe from the Hantaan gang of viruses.

In 1993, in the Four Corners, that comfortable notion came to an end. Fifty-four local residents, mostly Navajo Indians, came down with what at first seemed to be the flu—that catch-all conglomerate of symptoms that is both true flu and, too often, the prodromal opening round of so many other infectious diseases. In the Four Corners, the "flu's" shortness of breath worsened to become an acute respiratory distress. X-rays showed a picture of lung infiltrates, a waterlogged, drowning lung. The number of platelets (thrombocytes), the small blood cells involved in clotting, was markedly reduced. The heart and lungs failed in 40 percent of the patients and they died. Over that year, there were fifty-four cases of this new, bewildering disease. All had been healthy before their illness, even those who died. Their average age was thirty-four years.

A Hantaan-type virus was isolated from the Four Corners patients but its RNA sequence did not match that of any of the thirty or so known Euro-Asian species. This new American virus was named the Four Corners virus. However, according to the Scripps Research Institute's Dr. Michael B. A. Oldstone, this name was not good for business. The Four Corners economy largely depended on tourism. Having a virus and the deadly disease it caused named for the Four Corners was obviously not going to be a tourist attraction. So, according to Dr. Oldstone, after some political arm twisting the virus was renamed the Sin Nombre virus—which is Spanish for "no-named" virus. And who says scientists have no sense of humor?

Unlike the Euro-Asian hemorrhagic Hantaan infections, the Sin Nombre virus mainly targets the lungs, although subsequent clinical observations have found that hemorrhage and renal disease may accompany the pulmonary disease—the Hantavirus Pulmonary Syndrome. But as in all the Hantaan-group viruses, the main zoonotic reservoirs are rodents. Surveys of deer mice from Four Corners showed that about 10 percent carried the virus. The following winter was particularly harsh, and many deer mice took refuge in human habitations, garages, and storage sheds. When the human householders came in contact with the infected mouse droppings, the virus was transmitted to them.

Now that the Hantavirus Pulmonary Syndrome has been delineated as a clinical entity, it has been reported in some 300 cases in 31 states. However, the American Southwest remains the main focus of the disease with 73 percent of the infections coming from that region. Climate change has been but one factor in the Southwest outbreaks; what happened in Utah's Little Sahara Recreation Area revealed the syndrome's truly complex epidemiological mosaic. The Utah Bureau of Land Management has allowed the use of off-road recreational vehicles, those motorcycle-like, balloon-wheeled machines, in the Little Sahara. The vehicles destroyed much of the deer mouse's natural habitat and their population, already swollen by abnormally heavy rains, were forced into the tight corners of the remaining islands of habitat. There, they did what most overcrowded animals do—they fought and copiously defecated, transmitting the virus from mouse to mouse in blood, saliva, and feces. Human cases of Hanatavirus Pulmonary Syndrome were traced back to the Little Sahara Recreation Area. Mice were trapped and their blood analyzed for the virus to assess the level of zoonotic potential. The mice trapped in the off-road vehicle use area had four times the Sin Nombre infection rate than the mice trapped in the adjoining area where the machines were not permitted.

We have gone through most of the catalog of infectious disease categories that are, or could be, affected by global warming—vector-borne infections, mosquitoes (malaria and West Nile), ticks (TBE), zoonotic infections (Hantavirus, rodent reservoirs). There remains a last category, the waterborne infections, dominated by cholera, perhaps the most potentially dangerous and devastating of the climate-influenced diseases. Over centuries, its epidemics, the death by diarrhea, have been feared throughout the world. An 1832 cholera epidemic in Liverpool, England, drove its infuriated citizens to riot against the medical community who were helpless to aid the living or save the dying. Still, the Liverpudlians, with their inactive physicians, may have been better off than the Parisians who were also in the midst of a cholera outbreak. The leading Parisian doctor was François-Joseph-Victor Broussais, a "notorious bleeder," who believed that the anal outpouring could only be redressed by removing copious amounts of blood. London had the best of medical good fortune in Dr. John Snow, an early epidemiologist and the world's first anesthesiology specialist. In 1840, there was a severe cholera outbreak in London's St. James parish. Snow reasoned that it was a waterborne disease, probably traceable to the contamination from sewers into the drinking water supply. Few parish houses had a separate piped water system; people drew water from the communal pump at Broad Street. John Snow famously removed the pump handle and halted the epidemic.*

Finally, the father of bacteriology, Germany's Robert Koch, put cause and effect together with his 1883 discovery of the comma-shaped bacterium he named *Vibrio comma*, since renamed *Vibrio cholerae.* With identification of the responsible pathogen, the clinical and epidemiological picture cleared. Infection begins by ingesting the microbe in fecally contaminated food or water. After a short incubation period, less than a week, the fever and diarrhea begin and, untreated, gets worse and worse with a massive discharge of the "rice water" stool. Modern research has shown that the bacterium adheres to the intestinal lining by means of specific molecules. This is followed by the Vibrio's production of a very powerful toxin affecting the permeability of the intestine's cells. There is a massive excretion of water into the gut, and the body, losing so much fluid, goes into shock. About 50 percent of those infected and untreated, with the most virulent strains, die. These most virulent strains' genes transcribe the most potent toxin and other factors of intense pathogenicity. There are, however, other strains, such as the El Tor strain (named for the El Tor quarantine station in Saudi Arabia) that, while responsible for fever and diarrhea, are seldom lethal.

Now, no one need die, or even be long ill of cholera.

* There is now a pub in the parish (Gordon Square, I think) named the John Snow, where visiting public health professionals and microbiologists go for a memorial pint and sign the register maintained there. Snow might not have approved of the custom; he was a teetotaler.

Antibiotics such as ciprofloxicin and ampicillin can cure. One of the simplest ways to rescue the desperately diarrheic is by oral rehydration with a sugar-salt solution that restores both fluid and essential electrolytes.* Even so, the poorest of the poor, especially the Bangladeshis, remain at great risk to lethal cholera each monsoon season when the Brahmaputra River at its Bay of Bengal delta inundates the coastal villages. The few sewers back up and latrines overflow into the drinking-water wells. Cholera is the constantly recurring seasonal curse. Yet the Brahmaputra's largesse is so great that when the rains cease, villagers eagerly return to farm the renewed soils, even those on the river's islands. A prolonged monsoon means a prolonged cholera season, and this happened in 1991 and 1992 when El Niño extended and intensified the rains.†

Climate and cholera seemed to have a relatively, simple relationship—human feces in the rain-inundated sewer/well system. Then, Dr. Rita Colwell (now the director—she laughingly referred to her self as the Czarina—of the National Science Foundation) and her group made the complicating, disturbing discovery—cholera is, in a sense, a climate-

* Don't leave home without it! We always carry packets of rehydration mixture, and they have saved my wife and I when we have had, from Paris to Papua New Guinea, traveler's diarrhea. This is the formula, but packets can now be purchased as a commercial preparation: table salt 3.5 grams, sodium bicarbonate 2.5 grams, potassium chloride 1.5 grams, glucose 20 grams. Dissolve in a liter of boiled/bottled water. Rice powder can be substituted for glucose. You can also make a serviceable solution with 4 level tablespoons sugar, ¾ teaspoon table salt, 1 teaspoon baking soda, 1 cup orange juice, mixture brought to 1 liter with boiled/bottled water.

†The greatest choleregenic effect of that 1991 El Niño was not in Bangladesh but in Peru. It is believed that it began with the discharge of *Vibrio*-contaminated bilge water from a Chinese freighter. Cholera spread throughout the country and by the time it was controlled approximately 300,000 people had been infected. The death rate was 3 to 5 percent.

influenced vector-borne disease, the vectors being the creatures of plankton—copepods, algae, the larvae of myriad invertebrate species. *Vibrio cholerae* have adherent surface molecules which "glue" them to the plankton creatures, and there they survive over long periods of time. When the oceanic waters warm by aberrant climate changes, the costal waters become a soup of plankton *and* their *Vibrio cholerae* bacteria. Current work by NASA indicates that these plankton blooms can be detected by satellites and serve as predictors of where cholera outbreaks are most likely to occur.

Temporary climate changes undoubtedly affect the epidemiology of cholera and other waterborne infections. What is more uncertain are the effects of global warming. One prediction holds that, as the polar ice caps thaw, the sea level will rise. Sewers will back up in the inundated coastal regions. Drinking water will become contaminated, leading almost certainly to the spread of waterborne diarrheal diseases such as cholera. However, another group of experts assert that this is an alarmist prediction, that the earth will warm but slowly. Minnesotans will not wake up one fine morning to find themselves in the tropics. There will be plenty of time to apply our wealth and technology and create the infrastructure to meet the new challenges. What will happen to the peasant poor in the tropics, such as the Bangladeshis, is unpredictable. Their lot may be improved by simple effective steps now being put in place. Even passing drinking water through a piece of cloth to filter out the *Vibrio* and *Vibrio*-bearing plankton has reduced infection rates in the costal villages of Bangladesh. Vaccines, now of moderate or marginal value, will certainly be improved as the new, powerful immunological-genetic techniques are applied to the cholera and other waterborne pathogens. The organization Physicians for Social Responsibility claims that there will be "Death by Degree." Others believe that social and scientific responsibility will meet the medical challenges of global warming—that you really can't boil the lobster one degree at a time without its knowing it, and given the opportunity, climbing out of the pot.

Chapter

10

Loose Stools and Troubled Waters: Cryptosporidiosis

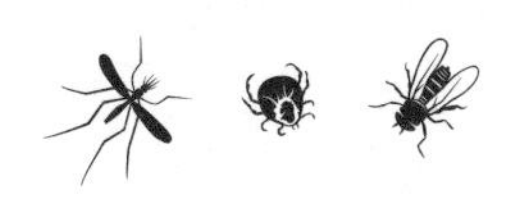

If you want an example of an emerging disease threatening America, forget Lassa or Ebola virus fevers tearing through some remote African village, and consider the *Cryptosporidium parvum* parasite on the diarrheic rampage in 403,000 Americans—killing some with AIDS and other severe immunodeficiency diseases. *Cryptosporidium*? Cryptosporidiosis? Before the massive Milwaukee outbreak in 1993, these were hardly "household" terms even among medical parasitologists or infectious disease specialists. I certainly never heard of it in my undergraduate and graduate parasitology courses, nor did I mention it when it eventually came my turn to deliver the lectures. It wasn't described in the medical parasitology texts or even in Tom Cheng's comprehensive, voluminous *General Parasitology* of 1973.

The first description of *Cryptosporidium parvum* goes back to 1912, ancient history by the present standards of scientific literature. In that year, a Harvard professor in the Department of Comparative Pathology, Ernest Edward Tyzzer, a physician who studied neglected protozoan parasites in neglected vertebrates, published a paper entitled, "*Cryptosporidium parvum* (new species), a coccidium found in the small intestine of the common mouse."* He named it *parvum*, "small," because of the small size of the "egg" oocyst, 4.5μ x 3μ (μ or μm is shorthand for micron, 1/1000th of a millimeter, or 0.000039 inch). In his experimentally infected mice, Tyzzer found large numbers of the parasite situated at the surface of the cells lining the small intestine. Despite the heavy load of parasites, there were no signs of illness.

Now to the Coccidia Chronicles: *Cryptosporidium* is a genus in the enormous Coccidia family. Some 400 species of Coccidia have been described as parasites of both invertebrate and vertebrate animals. The Coccidia have been mainly of interest to the veterinary parasitologist, because they cause great economic loss in domestic birds and mammals; to the human parasitologist, they have been only of "honorable mention" interest. Two species, *Isospora belli* and *Isospora hominis* infect humans. They are rarely diagnosed and cause a mild diarrhea lasting two or three days.

* I became acquainted with Tyzzer's work when I was a graduate student at the London School of Hygiene and Tropical Medicine. It was 1948; food—good food—was scarce in postwar England. My canny Scots professor, H. E. Shortt, convinced me that I should study a fascinating, lethal protozoan parasite of turkeys, *Histomonas meleagridis*. A large turkey run was built for me on the school's roof where I raised the turkey poults, mostly contributed by the commercial breeders in the nearby counties. From time to time, especially around Christmas, "H. E." would inquire about the status of the uninfected control birds. My rooftop turkey pen was next to the microbiology department's sheep pen (they used the sheep blood for culturing certain bacteria), and the morning sounds in London's Bloomsbury were resonant with gobblings and baa-ings. Tyzzer had carried out some of the early research on histomoniasis in 1919, and I referred to his published paper—and now do so again for his work on *Cryptosporidium parvum*—after an interval of fifty years.

The Coccidia are cousins of the malaria parasites; coccidiosis might be likened to a form of intestinal malaria. The two have a common basic life plan, each with alterations on the theme. The generations alternate, asexually reproducing then sexually reproducing from the union of "sperm" and "egg." The product within the fertilized "egg," the oocyst, are the infectious sporozoites. For the Plasmodium, this phase of the life cycle takes place in the mosquito; the oocyst with its maturing sporozoites attach to the mosquito's intestine. The sporozoites emerge to invade the salivary gland to await being transmitted when the mosquito takes her blood meal. For Coccidia, including *Cryptosporidium*, the oocyst forms in an epithelial cell lining the intestinal wall. It leaves the intestinal cell to be "layed" with the passing of the feces—into the sewer, your drinking water, swimming pool, or onto the feces-fertilized vegetable of your salad. When ingested, the sporozoites "hatch" from the oocyst and invade an epithelial cell to initiate the asexual reproductive cycle. As merozoites develop, more and more intestinal cells become parasitized. After several asexual generations, some of the merozoites are transformed into male and female stages—gametes. Fertilization takes place, and the oocyst is formed to pass out in the feces. And so coccidial life continues.

During the sixty-three years following Tyzzer's 1912 description of *Cryptosporidium* in the mouse, other species of the genus had been found in snakes, chickens, turkeys, guinea pigs, dingo dogs in Australia, jungle cats in India, and calves in America—but never in humans. The cryptosporidial separation of human and beast came to an end in 1975, with a severely diarrheic three-year-old girl from Tennessee.

The child's medical history was unremarkable—excellent

health with no allergies or indication of any other immune dysfunction. She lived on a farm in rural Tennessee with her parents, two cats, and a dog. The household water came from a well. Sometime in 1975, the child, normally bright and active, became lethargic and feverish, but the main trouble was a severe watery diarrhea. She had terrible cramps and vomited everything she ate. After enduring six days of acute illness, she was brought to Nashville's Vanderbilt University Hospital. X-rays and sigmoidoscopy showed a picture of an acutely inflamed large intestine. It was a colitis of unknown cause. All tests to identify a causative parasite or other microbial pathogen were negative.

Five days later, with her physicians still trying to determine a cause and course of treatment, the child took matters in her own hands and spontaneously began to recover. After nine days in the hospital, she returned home—stools now firm, no sign in X-rays or "scope" of any residual intestinal damage.

Hospitals move on expeditiously; the acutely sick and dying must be attended, work assignments efficiently and economically allocated. The recovered, discharged patients—particularly those that had self-recovered of such mundane illnesses as diarrhea become past history on the laboratory's diagnostic back burner. It was almost eighteen months after the child left the Nashville hospital that a pathologist examined the preserved rectal tissue biopsy sample. Under the light and electron microscopes, the pathologist saw a myriad of unknown organisms—round bodies attached to the border of the epithelial cells lining the intestine. With the assistance of veterinary pathologists, the X organism was identified as a *Cryptosporidium* of unknown species.

There was the expected modest flurry of excitement over a new, albeit nonlethal microbial parasite infecting a human. Eighteen months after the child's acute illness, medical investigators visited the farm. What was the source of infection? How did the child become infected? The parents were interviewed; the two, undoubtedly discomforted, cats had snippets of tissue taken from their rectum; the dog was gone. The epidemiological trail was cold. The cat tissues were negative for *Cryptosporidium.* The child had no history of contact with the farm's cattle. A case report was published in the April 1976 issue of the prestigious journal *Lancet.* The authors concluded with the indefinite statement, "the source of infection and mode of transmission to the child remains unidentified."

The annals of medicine are rich in exotic infectious oddities, most of them recorded, noted—and forgotten. But *Cryptosporidium* was about to emerge from its oddball status. The Tennessee doctors were unaware that about the same time they treated the *Cryptosporidium* child, on the other side of America, in Seattle, a thirty-nine-year-old immunosuppressed man had developed an overwhelming diarrhea caused by the same organism.

The man was a victim of his own immune system. At twenty-two years of age, ulcerative colitis attacked his gut and the offending bowel was surgically removed. He did well for the next seventeen years until, in June 1974, the perfidious immune system targeted his skin with bullous pemphigoid. These are great blisters an inch or more in diameter, usually concentrated in areas where the body flexes. Secondary infection can be a serious complication. There is no cure for pemphigoid, but since all evidence points to autoimmunity, it is treated with immune suppressing drugs. Throughout that summer, the man was given heroic doses of cyclophosphamide, vincristine, and prednisolone. By October 4, he was well enough to do some construction work on his home—a farm. The next day, he had abdominal cramps and the day after that, October 6, he had a massive watery diarrhea so severe that he had to return to hospital.

Culture and examination of the fluid draining from his ileostomy for the usual suspects—Shigella and Salmonella bacteria, protozoan and helminth intestinal parasites—were negative. A biopsy was taken from the remaining small intestine. He was put on a fluid and diet regimen similar to that given to patients with cholera and sprue. Two weeks later, the diarrhea disappeared and he gradually returned to full health.

The biopsy tissue taken during the acute phase was grossly abnormal. The villi—those microscopic "fingers" absorbing nutrients—were either shortened, or just gone. And there they were under the microscope—the small, spherical mystery bodies staining bright blue with red centers. Electron microscopy revealed that the bodies covered by the host cell's membrane at the outer border were trophozoites, gametes and oocysts of *Cryptosporidium.* For this second case, there also was an epidemiologic follow up and again it was noncontributory. He lived on a farm in rural Washington. There were beef cattle and a dog. Field mice and deer were often seen on the farm. No one else in his family had fallen ill with diarrhea.

Two cases don't make a health threat or even a "textbook" disease category. One was a child, the other an immunocompromised adult susceptible to all sorts of "abnormal" opportunistic pathogens—epidemiological freaks of nature. In 1976, few microbiologists envisioned that *Cryptosporidium* would become a major global player in the diarrhea sweepstakes.

Fevers of unknown origin and diarrheas of unknown origin are diagnostic failures—but not from want of trying. The pathogen may be either unrecognized or unfamiliar to both clinician and laboratory technologist. The specific techniques and reagents may be unavailable to the clinical laboratories. Most patients in diagnostic limbo recover with the help of supportive treatment. No great effort would be made to identify the causative pathogen. In the 1980s, such indifference ended when AIDS came to our countrymen. Intractable, incurable diarrhea was frequently the messenger of impending, full-blown AIDS. In Africa, diarrhea so emaciated its victims that AIDS was known as the "slim disease."

An undefined etiology for the AIDS-associated diarrheas was unacceptable; laboratory diagnosis was pursued more aggressively than for "normal" diarrheas. When nothing else turned up, the *Cryptosporidium* in the immunocompromised man came to mind. The difficulty lay in diagnosing *Cryptosporidium* by the customary techniques applied to other intestinal parasitic protozoa, such as *Entamoeba histolytica* (amebiasis) and *Giardia lamblia* (giardiasis). New techniques were devised to concentrate and stain the *Cryptosporidium* in the fecal specimen. When these methodologies were applied, laboratory personnel came to recognize that those little red dots in the stained specimen seen under the microscope were the oocysts of *Cryptosporidium.*

Thousands of AIDS-associated diarrheic stools were examined for *Cryptosporidium.* When the results were tabulated, over 10 percent of the specimens were positive. Cryptosporidiosis was obviously an important cause of illness in people with HIV, but there was a perplexing inconsistency in its severity. For some, it was a transient diarrhea that ended in self-cure—similar to the course in the HIV-free immunocompetent. In a second group, the infection either interminably persisted or took a much longer time until self-cure intervened. The moderately severe diarrhea was accompanied by an abnormal histological architecture of the intestinal mucosa. The sickest, third group suffered from a savage life-threatening diarrhea. In these cases, the parasites were often disseminated beyond the intestinal tract into the bile duct and, in a few cases, the respiratory tract. Biliary infection caused such intense abdominal pain that surgical removal of the bile duct was necessary. Even this desperate measure did not bring relief to all those affected. Within a year of diagnosis, about half the AIDS patients with severe cryptosporidiosis were dead.

Now that diagnostic methods were in place, physicians, clinical laboratories, and infectious disease investigators began considering the possibility of *Cryptosporidium* as a cause of diarrhea in the normal immunocompetent public. The CDC made cryptosporidiosis a reportable disease. By the mid-1990s, the results of this broad fecal search revealed that *Cryptosporidium* is a common cause of diarrhea, especially in the children of the third world. In the United States 2,000 to 3,000 cases were reported to the CDC each year. It occurred in virtually every state in the nation. Moreover, it was suspected that many infected people were asymptomatic, but nevertheless passing oocysts—the Typhoid Mary phenomenon. Overall, stool surveys revealed that there were many *Cryptosporidium* Marys; as many as 10 percent in some areas of the United States. The essential epidemiological and clinical aspects were now understood—the widespread commonness of infection and the wide spectrum of clinical expression. However, even in 1980, more was known about the disease than the parasite causing it. What was it species-wise? Was it zoonotic—and if so, what animals are naturally infected? Importantly, how do humans acquire the infections?

The *Cryptosporidium* in humans shared a common characteristic: the oocysts shed in the stool were small—very small—5μ (compared to a red blood cell which is 7μ in diameter). The only described species with an oocyst that small was Tyzzer's 1912 *Cryptosporidium parvum* in the common house mouse. By 1985, papers on cryptosporidiosis in humans were referring to the causative species as *parvum.* The full range of its infectivity wasn't known, but, it was assumed from the 1912 study, that it was sort of monogamous, faithful only to its mouse host. If the species was truly *parvum* it would mean that cryptosporidiosis was a zoonoses with a rodent reservoir. But was human and mouse the full host range?

It was reminiscent of New York's West Nile, in which case veterinarians and physicians were dealing with two diseases that were the same disease. Thirty years ago, veterinarians in America and Europe were called by worried dairy farmers to examine their very sick, diarrheic calves. Newborn calves were most affected, some died. None of the "standard" intestinal pathogens could be isolated from these animals. Instead, they found oocysts of a size that pointed to *Cryptosporidium parvum.*

In the 1980s, infectious disease investigators were wondering whether the *parvum* in diarrheic humans was a zoonosis, while veterinary pathologists were wondering whether the *parvum* in diarrheic calves was a zoonosis. Cross-infection experiments carried out in the Netherlands, Italy, and the

United States over a ten-year period proved that *Cryptosporidium parvum* had host flexibility. Cows, mice, and humans all had the same pathogen, capable of infecting each other and one another. Nor did it require any great number of those 5-μ oocysts to initiate infection. Cynthia Chappell and her colleagues at the University of Texas School of Public Health in Houston gave graded doses of *Cryptosporidium parvum* from a calf to healthy human volunteers. Seventy-one percent of those given the highest number, 1,000 oocysts, became infected and experienced the transient diarrhea, nausea, and cramping. However, 20 percent of those given the lowest dose, a mere 30 oocysts, also became infected.

The mice may have started it all. In one survey in the United States, half the mice trapped were infected and shedding oocysts. Mice may be only incidental epidemiological contributors; the asymptomatic cows and humans excreting millions of oocysts each day are perfectly capable of perpetuating the infection cycle by contamination. A cow munching on grass on which another, infected, cow had dropped its dung would, in turn become infected. Farm dwellers—the Tennessee child and the Washington man—can also become infected by contaminative mischance; after all, it only takes a dab—30 oocysts—to do you. A hospital diarrheic intern showed how readily human to human transmission can take place. He was responsible for the care of a patient with chronic cryptosporidiosis. All the sanitary procedures laid down in dealing with infectious disease patients were observed. Nevertheless, within two weeks of first contact, the intern had loose stools and abdominal cramps accompanied by the shedding of oocysts. However, all these infections seemed to be derived from chance exposure to contaminated material. There was, as yet no sign of broad environmental involvement leading to a major outbreak—an air filled with suspended oocysts, a food supply at supermarkets contaminated with oocysts, and most importantly, a drinking water supply with oocysts.

The first example of an environmental source came, in 1984, from a Texas suburban community of 1,800 households. An unaccustomed number of people with diarrhea were sitting, squirming (ten or more watery, explosive stools per day) in the local doctors' waiting rooms. The Health Department eventually became aware of the outbreak and conducted a telephone survey which revealed that 34 percent of the community had had gastroenteritis, lasting several days to a week, during that past month. A waterborne pathogen was suspected, but the community water supply came from an artesian well—a water source, touted by bottled water companies as insuring immaculate purity. Not this Texas artesian—tests showed that its purity was fouled by feces. New stool examinations and serologic tests indicated *Cryptosporidium* as the gastroenteritis-causing pathogen in the drinking water. The report on the outbreak, published in 1985, came to the disquieting conclusion that "*Cryptosporidium* should be added to the list of waterborne organisms causing outbreaks of gastroenteritis." Still, it was a small outbreak in a small Texas town. There was no great national concern. That was to come with the Great 1992 Milwaukee Cryptosporidium Horror Show. It was to be the largest documented outbreak of a waterborne disease ever to have occurred in the United States.

Spring rains along Lake Michigan were said to have been especially heavy in 1993. Milwaukee, Wisconsin, a city of about 650,000 (a million more if the environs are included), is on the lake's southwestern shore—not all that far from Chicago. Water to the city and its outskirts comes from the lake after its treatment by the Milwaukee Water Works where it is treated by chlorination, clumping of suspended particles with a coagulant, and, finally, passed through a sand filter. What comes out of the tap may not be Vitel but it had always been clear and safe—until March 1983, when the water turned murky and the stools watery. Its exact day of beginning is uncertain. Later interviews point to the first week of the month, because, by March 14, 2,300 fecal specimens had been sent by doctors to fourteen clinical laboratories for routine testing of bacterial, parasitic, and viral intestinal pathogens. There were a few positives, the odd shigella and giardia, but nothing to account for the mass diarrhea gripping Milwaukee. There had been no special examinations for *Cryptosporidium.*

The public health agencies of city and state responsible for investigation and defense remained unaware of the fecal explosion. The doctors, looking at the negative lab reports, didn't find cause to notify the authorities. Milwaukee's city health department became alert that something was amiss by the extraordinary absenteeism in the city departments, schools, and hospital staff. On April 5, approximately one month after the first diarrheas, the city health department notified the Wisconsin Division of Health that not all was well in Milwaukee. Two days later, almost serendipitously, two of the clinical laboratories had the forethought and expertise to test for *Cryptosporidium* in seven stool specimens sent to them. All were positive for the parasite. The health department moved swiftly and decisively.

Experience led to the working hypothesis that a gastroenteritis outbreak of this dimension probably sprang from a contaminated water source. The water purification plants were inspected and found at fault. *Cryptosporidium* oocysts were found in the water emerging from the sand filter and ready to go into the mains. Ice blocks frozen on March 25 by a Milwaukee company were melted. The samples contained oocysts. Clearly, purification procedures were not keeping oocysts out of the water system. On April 9, the water works in south Milwaukee was closed down until safer methods could be implemented. People were advised to boil their drinking water or to exclusively use bottled water.

Clinical laboratories were directed to test all stool specimens for the presence of *Cryptosporidium* oocysts. Within two months, to May 15, 739 people had been diagnosed with active cryptosporidiosis. After looking at that figure, the health department hit the phones. Interviewers asked if the person called or a member of the family had diarrhea and if so when? For how long? Was there fever? Weight loss? How many times a day did they have to go to the toilet? Were they HIV-positive or taking any anticancer or other immunosuppressive drugs?

When all the information was collected and collated, the picture of a massive outbreak emerged. An estimated 403,000 people had been infected and affected. The diarrhea lasted an average of nine days during which they passed a watery stool an average of ten times a day (but up to a devastating ninety daily toilet calls). They had fever, averaging 101°F, and lost an average of ten pounds.

Milwaukee had firmly fixed *Cryptosporidium* in the pathogen pantheon. Lessons were learned but mysteries remained. The big enigma was uncertainty over the parasite's origins. The turbidity of the tap water indicated that something was amiss in the water works. Oddly, however, the drinking water had begun to cloud on March 23, while the diarrheas had begun several weeks earlier. The best guess had it that runoff from dairy farms and slaughterhouses into the rain-swollen rivers entering Lake Michigan were the source of the oocysts, although contamination from human sewage couldn't be discounted. From cow or human, the *Cryptosporidium* oocyst turned out to be a tenacious survivor, killed neither by chlorination nor other standard purification methods. Back-flushing to clean the sand filter actually concentrated the oocysts. Especially worrisome was the great vulnerability of the HIV-positive and other immunosuppressed to remorseless, deadly cryptosporidiosis. A lesson that may still not have been learned and acted on was the dissociation between practicing physicians and public health authorities. It took too long before the city and health departments were alerted and activated.

Spring gave way to summer in Milwaukee; rivers receded, municipal water works were carefully monitored, slaughterhouses admonished, and the Great 1993 Milwaukee *Cryptosporidium* Outbreak departed the intestinally challenged city.

During the decade since Milwaukee's *Cryptosporidium parvum* epidemic, there have been other outbreaks of cryptosporidium gastroenteritis—although none so large as Milwaukee's. There have been human-to-human transmissions, especially among the HIV-positive. There has been transmission from "recreational waters"—investigators were surprised to find how often children (and some adults) defecate in swimming pools. However, in every case, the major sources of infection have, once more, been traced to the trusted municipal water. Surveys have revealed that anywhere from 65 to 97 percent of all surface waters in the United States are contaminated with *Cryptosporidium* oocysts!

And there is not a great deal that the cities can do to clean up their drinking water. The present municipal water disinfection methods just simply don't work. Water coming from the faucet may well meet all the current federal standards of purity and still give you cryptosporidiosis. In 1994, there was a Cryptosporidium outbreak in Clark County, Nevada. Dr. Susan T. Goldstein and her team of epidemiologists described the outbreak in a paper despairingly called: "Cryptosporidiosis: An outbreak associated with drinking water despite state-of-the-art water treatment."

New methods to exclude the oocysts and new compounds to kill them are being explored. They may, or may not, work; even if they do work, they may not be implemented. It won't be cheap to change the workings of a water works. Cutting off the oocysts at their animal source will also be essential. It will be necessary to confront our delicate reluctance to contemplate the fecal facts of life—domestic animals, the zoonotic source of *Cryptosporidium,* produce millions of tons of droppings daily, and you can't legislate against physiology. Fecal material, including the oocysts, inevitably seeps into the surface waters. According to the scenario by the Carnegie Mellon University's Elizabeth Casman and Hadi Dowlatabadi, global warming will bring more and more contaminative seepage from the increased rain and swollen rivers.

When I lived in Northern Nigeria's Bauchi Plateau in the 1950s, it came to my attention that our major domo, Amadou, was supplementing his salary by selling the family feces. These came from the septic pit and were marketed to pagan farmers who believed that the white man's protein-rich diet made his feces a superior fertilizer. *Cryptosporidium* and the turbulence of international events may force Americans to think like Nigerian pagans—that feces are too valuable to waste. The technology for extracting energy (and inactivating oocysts in the process)—especially methane—from feces is already available. We only need the will and acceptance of transitional trade-offs (say a little SUV for a big SUV) to exploit this bounteous natural gift. And wouldn't it be an ironical, symbolic triumph if our liberation from oil came from the bovine bowel.

Index

Index

Index

Index